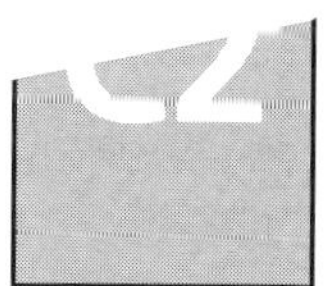

Practice for Book C2

Contents

PUBLISHED BY THE PRESS SYNDICATE OF THE UNIVERSITY OF CAMBRIDGE
The Pitt Building, Trumpington Street, Cambridge, United Kingdom

CAMBRIDGE UNIVERSITY PRESS
The Edinburgh Building, Cambridge CB2 2RU, UK
40 West 20th Street, New York, NY 10011-4211, USA
477 Williamstown Road, Port Melbourne, VIC 3207, Australia
Ruiz de Alarcón 13, 28014 Madrid, Spain
Dock House, The Waterfront, Cape Town 8001, South Africa

http://www.cambridge.org

First published 2001

Printed in the United Kingdom at the University Press, Cambridge
Typeface Minion *System* QuarkXPress®

A catalogue record for this book is available from the British Library

ISBN 0 521 79869 8 paperback

Typesetting and technical illustrations by The School Mathematics Project
Illustrations on pages 18, 19, 21, 47 and 68 by Chris Evans
Cover image © Image Bank/Antonio Rosario
Cover design by Angela Ashton

1 Graphs that tell stories

Section A

1 Ffion is timing a hotel lift as it moves between floors.

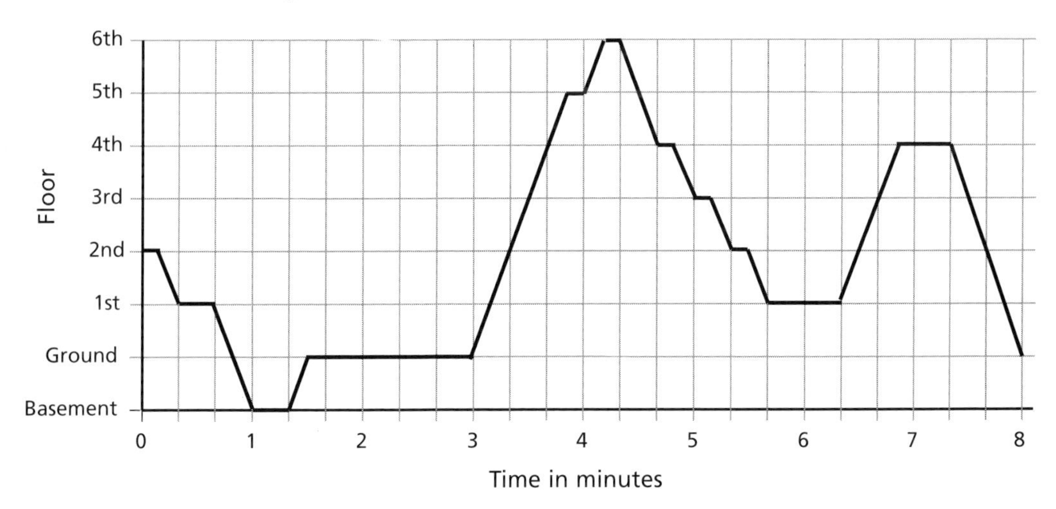

When she starts timing, at 0 minutes, the lift is at the 2nd floor.

(a) Write down what the lift does in the first minute.

(b) What do you think happens at 3 minutes?

(c) How long is the lift at the 5th floor?

(d) Sam is on the 3rd floor.
At 3 minutes he pushes the button for the lift.
How long does he have to wait for the lift to arrive?

(e) During the 8 minutes shown in the graph, Gill takes the lift up to the 4th floor.
At what floor do you think she gets in the lift?

(f) For how long during these 8 minutes is the lift above the level of the 3rd floor?

2 Write a story for this graph.

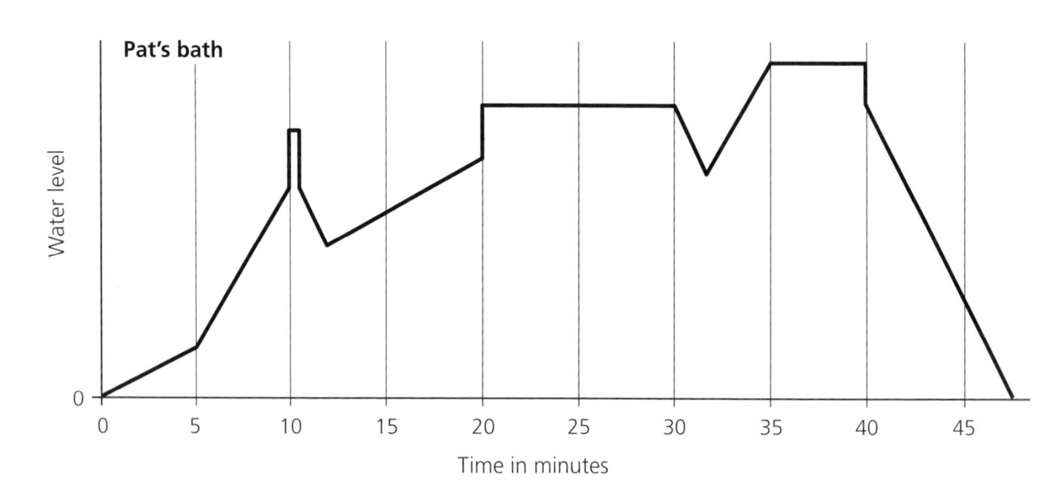

3 This graph shows the number of people in a shopping centre one day near Christmas.

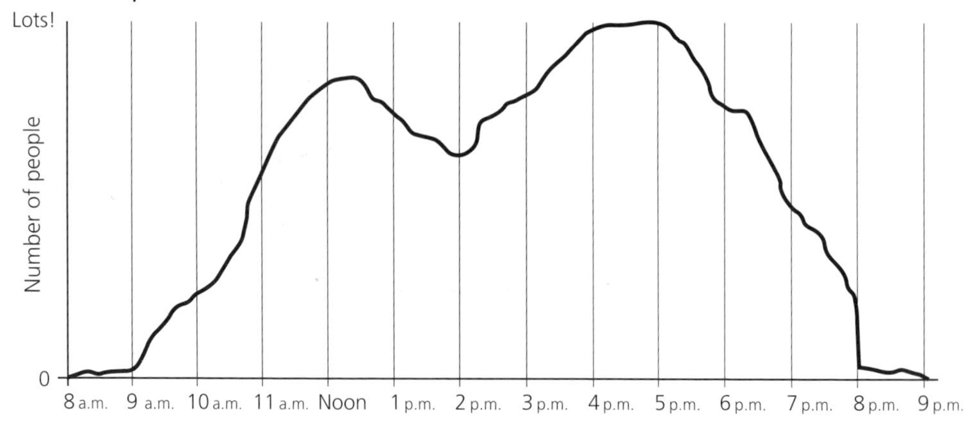

(a) At what time do you think the shopping centre opened? 8:00 Am

(b) What time do you think it closed? 9 pm

(c) At roughly what time were there most people in the centre? 5:Pm

(d) Were there more people in the centre at 3 p.m. or 4 p.m.? More at 4

(e) At 11 a.m., were more people coming in to the centre or leaving it? Coming in

(f) At 1 p.m., were more people coming in or leaving? Leaving

*4 In a shop there is only one checkout. The graph shows the number of people in the queue, including the person being served.

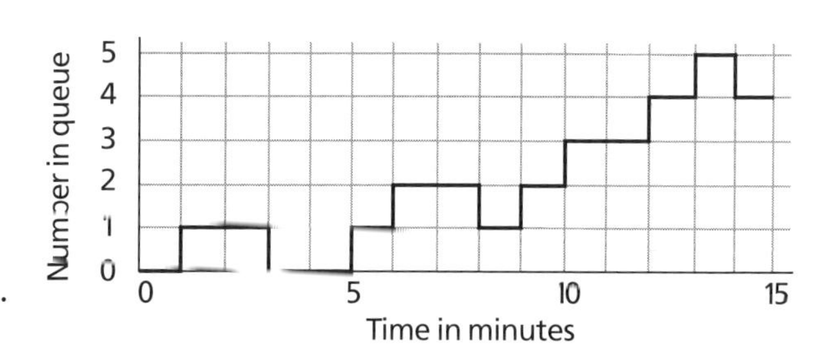

Nobody leaves the queue without being served.

(a) The first person to be served went to the checkout at 1 minute. When did he leave?

(b) When did the second person go to the checkout?

(c) When did the second person leave?

(d) When did the third person (i) join the queue (ii) leave the checkout?

(e) How many people were there in the queue after the third person left?

Section B

1 Rajan has a glass of cola. He is drinking the cola with a straw.

He sucks the cola out of the glass at a steady rate, without stopping for breath!

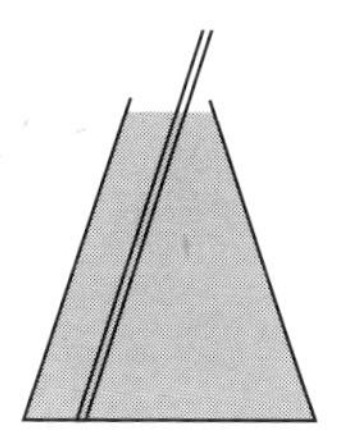

Which graph shows the level of the cola in the glass as it empties?

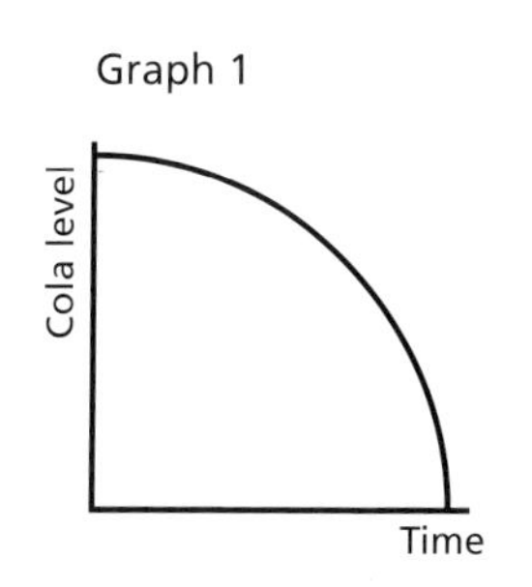

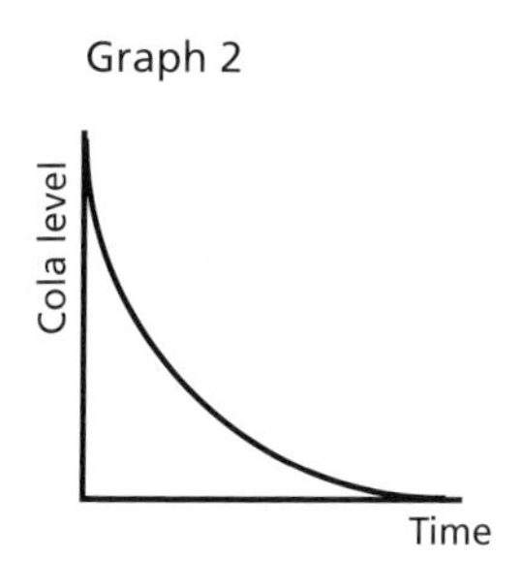

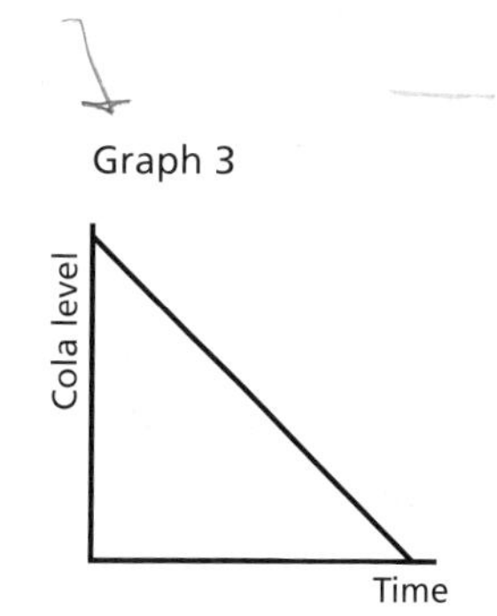

2 Here are three glasses and three graphs.
Each glass is emptied steadily.
Which graph goes with which glass?

A B C

Graph 1 — Cola level / Time

Graph 2 — Cola level / Time

Graph 3 — Cola level / Time

3 Imagine each of these glasses is emptied steadily.
Draw sketch graphs of how the cola level changes.

A B C D

4 This graph shows the level of water in a bottle as it steadily empties.
Draw a bottle which fits the graph.

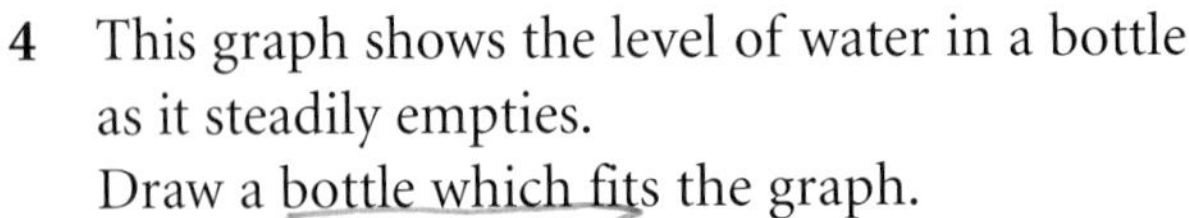

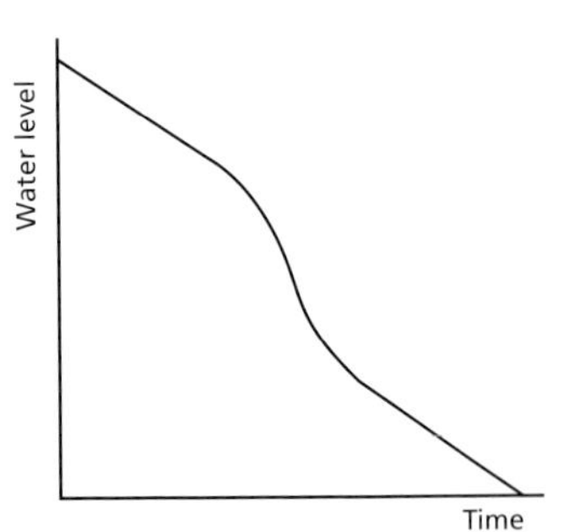

Section C

1 Cass went for a cycle ride.
This graph shows her speed.

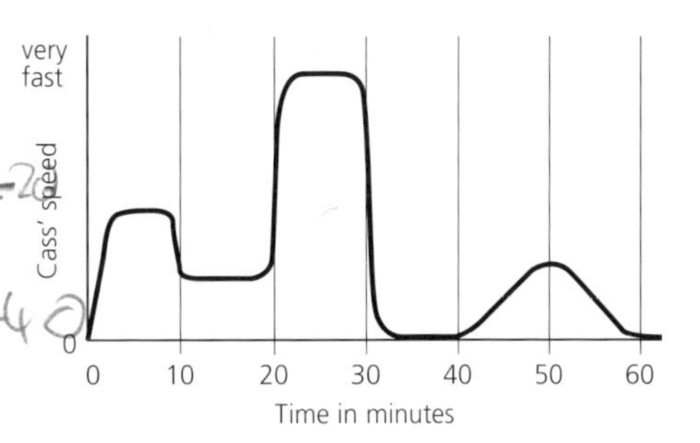

(a) She stopped for a few minutes.
Between what times was this?

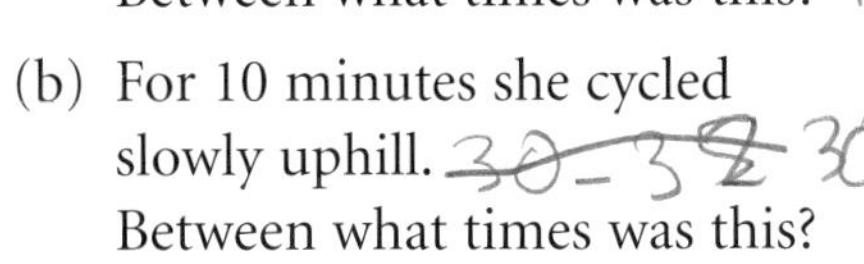

(b) For 10 minutes she cycled slowly uphill.
Between what times was this?

(c) For 10 minutes she cycled quickly downhill.
Between what times was this?

2 Jaz wrote this in his diary.

Sketch a graph of his speed on the ride.

Label the up-axis 'Speed' with '0' at the bottom and 'very fast' at the top.

> *Went for a ride – a whole hour!*
>
> *First 10 minutes were OK – went quite fast.*
>
> *But then it took me 10 whole minutes to get up Tunnel Hill, going very slowly.*
> *I had to stop for 5 minutes at the top.*
> *But then only 5 minutes very fast down the other side.*
>
> *Met H and pushed the bike for 15 minutes, then stopped at the lake for 5 minutes.*
>
> *Then it took me 10 minutes to get home – quite fast again.*

3 A man competes in a sports event.
This graph shows his speed.

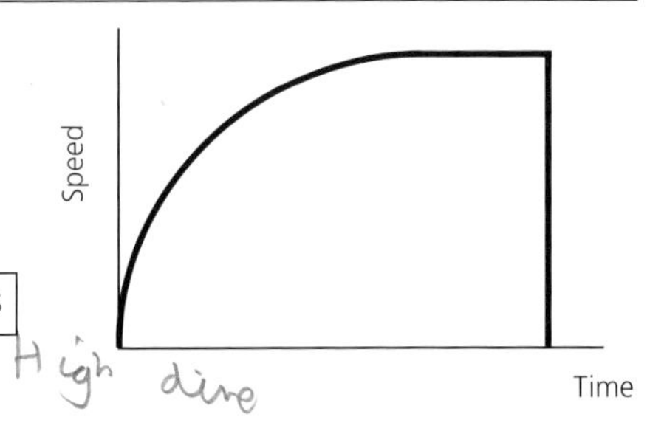

(a) In which of these sports do you think he was competing?

Sprint | Long jump | High dive | Darts

(b) Think about the other three sports.
For each of them, draw a sketch graph of the competitor's speed.

Explain how each sport fits its graph.

2 Scaling

Sections A, B and C

1 Copy this shape on to squared paper.

Now draw an enlargement of it using scale factor 2.

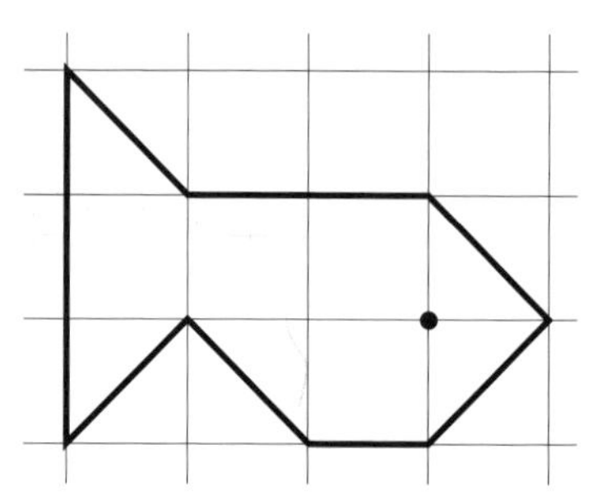

2 (a) What can you say about the angles in these two shapes?

(b) Is the larger shape an enlargement of the smaller one?

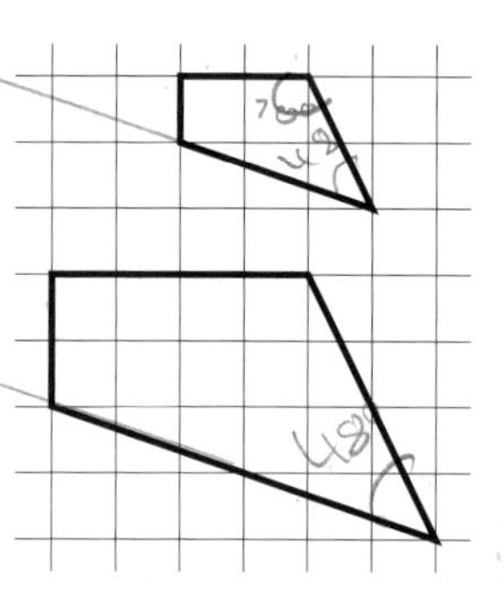

3 (a) What can you say about the angles in these two shapes?

(b) Is the larger shape an enlargement of the smaller one?

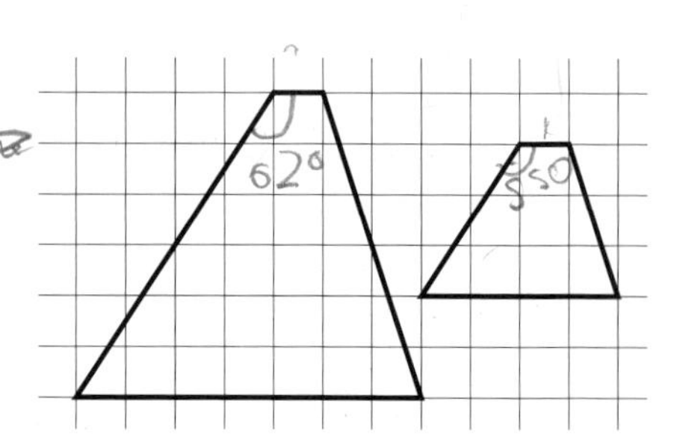

4 Which shapes are enlargements of which? Give the scale factors.

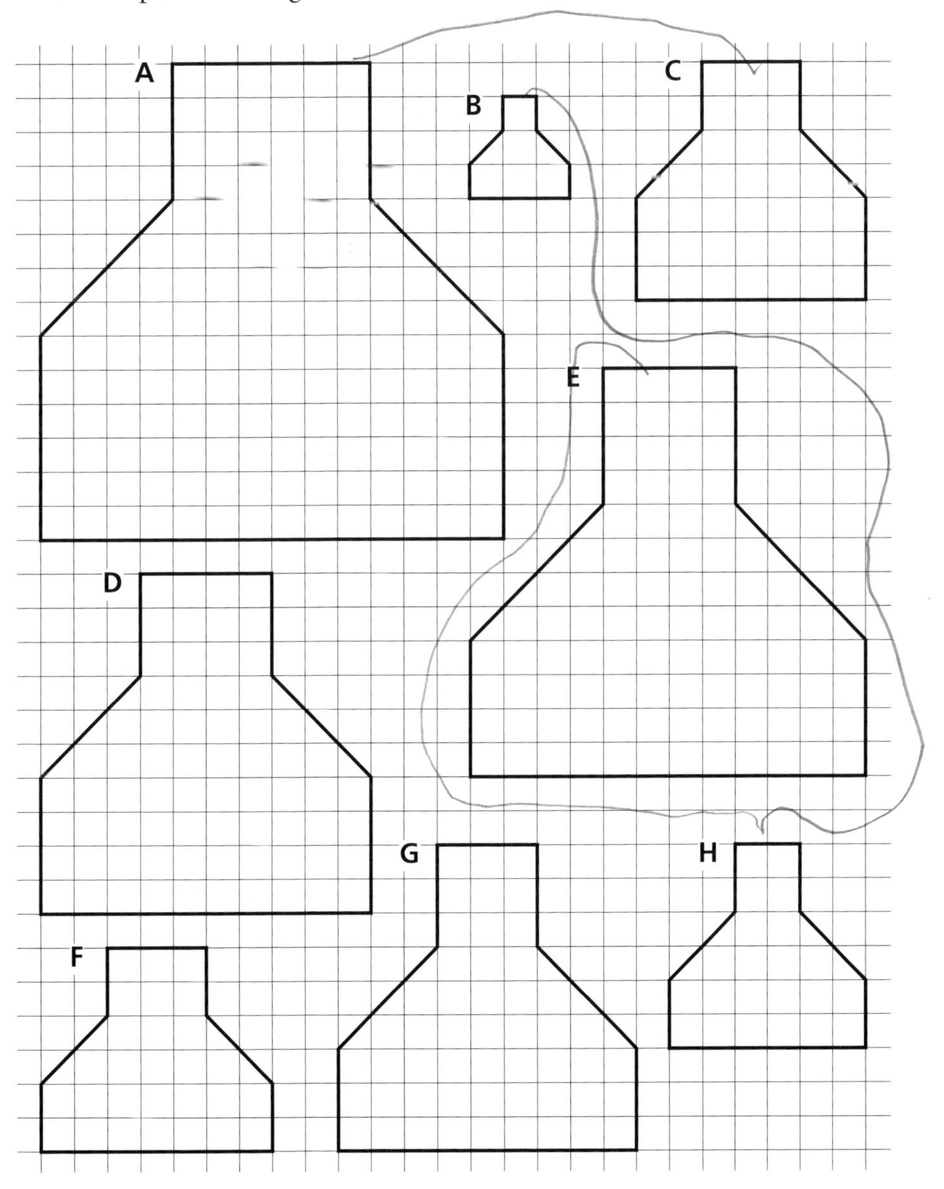

5 These are sketches of some right-angled triangles. They are not drawn accurately.

Which triangles are enlargements of which?

Give the scale factors.

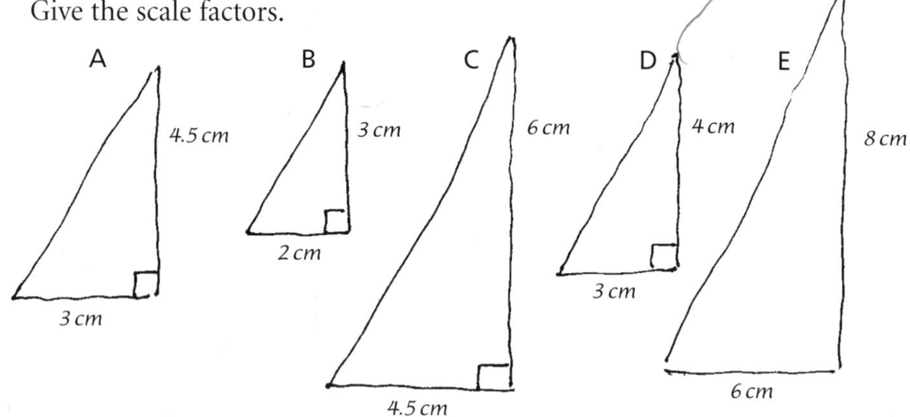

6 What is the scale factor of each of these?

(a) A scaling from P to Q

(b) A scaling from Q to P

P  Q 

7 What is the scale factor of each of these?

(a) A scaling from R to S

(b) A scaling from S to R

R  S 

Section D

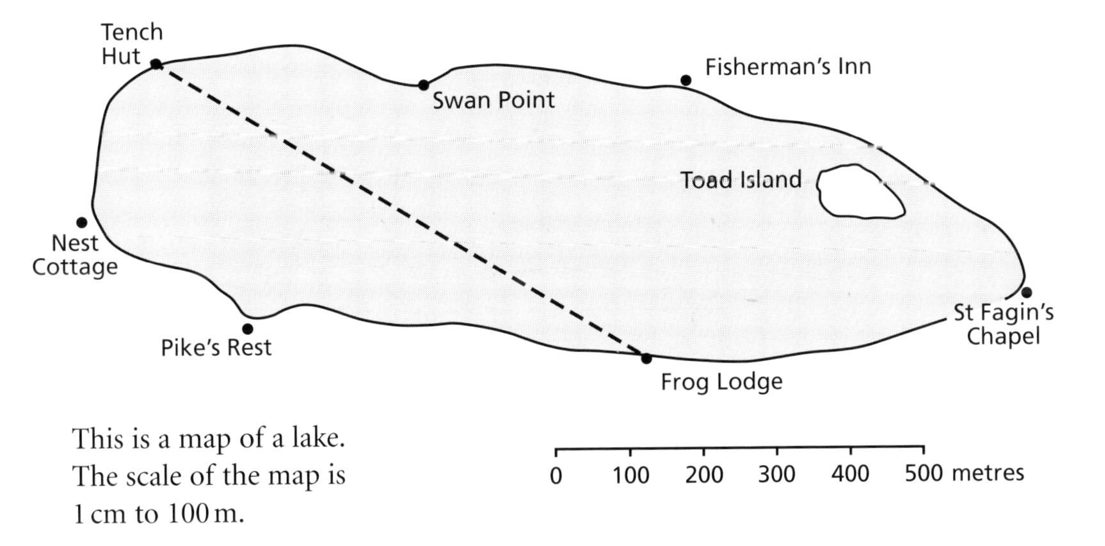

This is a map of a lake.
The scale of the map is
1 cm to 100 m.

1 (a) Measure the distance in centimetres on the map between Tench Hut and Frog Lodge.

(b) What is the actual distance, in metres, between Tench Hut and Frog Lodge?

2 Find these actual distances, in metres.

(a) From Pike's Rest to Fisherman's Inn

(b) From Nest Cottage to Swan Point

(c) From Nest Cottage to St Fagin's Chapel

(d) From Fisherman's Inn to Frog Lodge

3 The edge of the lake is curved.
Describe how you could estimate from the map the perimeter of the lake.

4 This is a map of Toad Island, drawn to a larger scale.
1 cm represents 20 m.

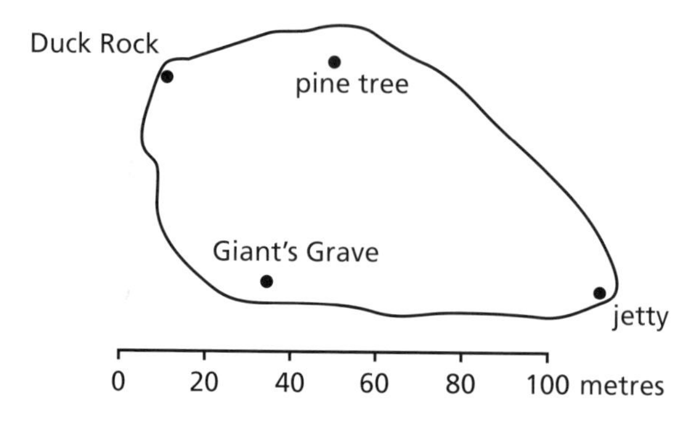

Find the actual distance, in metres, between each of these places.

(a) Duck Rock and the jetty

(b) Giant's Grave and the pine tree

(c) Giant's Grave and the jetty

5 The map of Toad Island in question 4 is an enlargement of the map on page 11.

What is the scale factor of the enlargement?

6 What will these measurements of the Humber Bridge be, drawn to a scale where 1 cm represents 100 metres?

(a) The 1420 metre length of the central span

(b) The 290 metre height of the towers

Section E

1 Write each of these map scales as a ratio.

(a) 1 cm to 100 m

(b) 1 cm to 250 m

(c) 1 cm to 5 km

(d) 1 cm to 1 m

2 Write each of these scales as a ratio.

(a) 2 cm to 1 km

(b) 2 cm to 50 m

(c) 5 cm to 100 m

3 Graphs and charts

Section A

1 The graph below shows the temperatures in °C in three cities. It shows the temperatures at noon in Sydney, London and Madrid.

The graph covers 15 days in May.

Temperatures in three cities

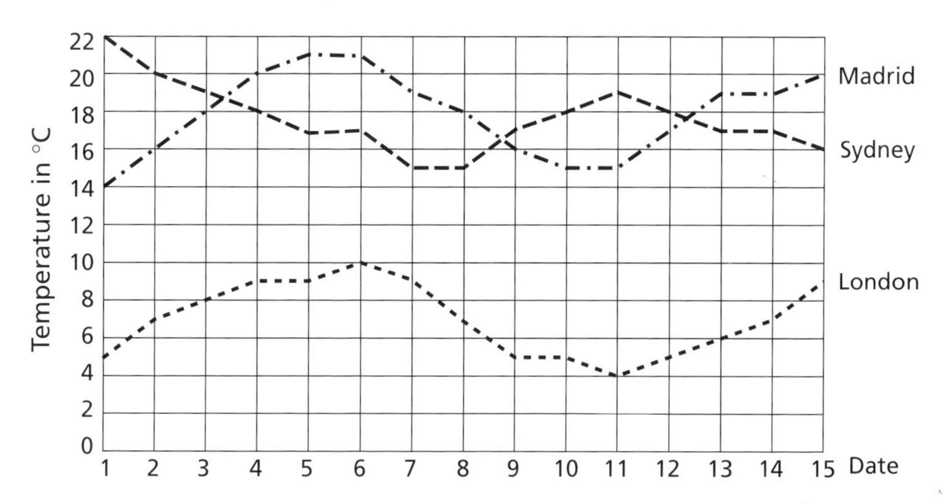

(a) Which city had the lowest noon temperature? What was it?

(b) Which city had the highest noon temperature? What was it?

(c) The noon temperature in London went up, then fell to its lowest on the 11th May, and then went up again. Describe what happened to the temperature in Sydney.

(d) On which dates was it hotter at noon in Madrid than in Sydney?

(e) On what date was there the greatest difference between the noon temperature in Madrid and in Sydney? What was the difference?

(f) What is the highest temperature in London shown on the graph?

(g) On how many days was the noon temperature in London 8°C or above?

Section D

Represent each of these sets of data graphically in a suitable way.

1 These figures show the weights of 50 fish caught in a pond.
The weights are all in grams.

17	56	34	18	19	54	8	45	43	21
33	38	32	21	41	36	32	28	29	30
42	51	18	55	12	10	40	39	27	44
17	54	34	24	48	59	32	50	42	43
9	43	38	35	55	12	42	48	49	41

2 This table shows the rainfall each month in Miami one year.
It shows how much rain fell each month in millimetres.

Jan	Feb	Mar	Apr	May	Jun	Jul	Aug	Sep	Oct	Nov	Dec
58	52	56	65	185	195	137	175	215	220	95	45

3 This table shows the amount of homework done by pupils in Y8 and Y10 one night.

The figures shown are the percentages of each year-group.

Time (minutes)	Y8	Y10
None	18	12
1 to 30	39	30
30 to 60	35	42
60 to 90	7	12
90 to 120	3	4
over 120	2	0

4 Solving equations

Section A

1 Solve each of these equations.
Show your working.

(a) $3p + 4 = 31$

(b) $15 + 4n = 51$

(c) $c + 11 = 5c + 1$

(d) $3x + 7 = 4x + 4$

(e) $7y + 17 = 23 + 4y$

(f) $15 + 6a = 2a + 23$

(g) $3.5 + 2m = 2 + 5m$

(h) $10 + 0.5w = 4w + 3$

(i) $7 + 0.2h = h + 3$

(j) $\frac{d}{4} + 32 = d + 2$

2 Make up an equation where the solution is $x = 7$.
Check your equation works.

Section B

1 Turn each of these number puzzles into equations and solve them.

(a)

Nick thinks of a number.
He multiplies it by 7.
Then he adds on 21.
His answer is 10 times as big as his starting number.

(b)

Cassie thinks of a number.
She adds on 4.
Then she multiplies by 5.
Her answer is 50.

(c)

Polly thinks of a number.
She adds 2.
Then she multiples by 3.
She adds on 10.
Her answer is 5 times as big as her starting number.

(d)

Jack thinks of a number.
He halves it.
He adds on 10.
His answer is 3 times as big as his starting number.

2 Meg and Sean both think of the same number.
Meg multiplies her number by 6 and adds 8.
Sean multiplies his number by 4 and adds 20.
They both get the same answer.

What was their starting number?

3 Paul and Jane start with the same number.
Paul adds 12. Jane adds 2.
Paul's answer is three times Jane's answer.

What was their starting number?

Section C

1 Solve these equations. Show your working and check your answers.

(a) $7n - 2 = 4n + 7$

(b) $b - 3 = 3b - 11$

(c) $3y + 12 = 7y - 4$

(d) $11f - 17 = 10f + 19$

(e) $x - 25 = 4x - 61$

(f) $3p - 2 = 1 + p$

(g) $\frac{m}{2} + 10 = m - 1$

(h) $0.2w - 2 = 0.1w - 1$

(i) $2(x - 2) = x + 8$

(j) $3y + 10 = 4(y - 1)$

(k) $6(a + 1) = 4(a + 4)$

(l) $3(b + 2) - b = 8b - 18$

2 Pat and Peg both think of the same number.
Pat subtracts 1, then multiplies by 4.
Peg adds 5, then multiplies by 2.
They both get the same answer.

What number were they thinking of?

3 Carl drew two rectangles. The first rectangle was 4 cm wide.
The second rectangle was 6 cm wide, and its length
was 3 cm less than the length of the first rectangle.
The area of the second rectangle was 2 cm^2 greater
than the area of the first rectangle.

How long was each rectangle?

4 These triangles both have the same area.
Find x.

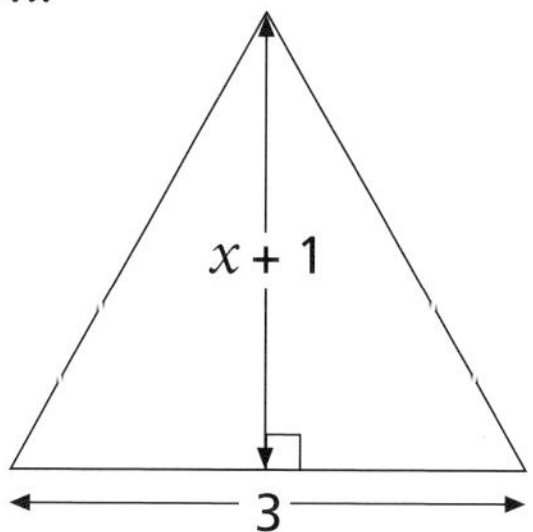

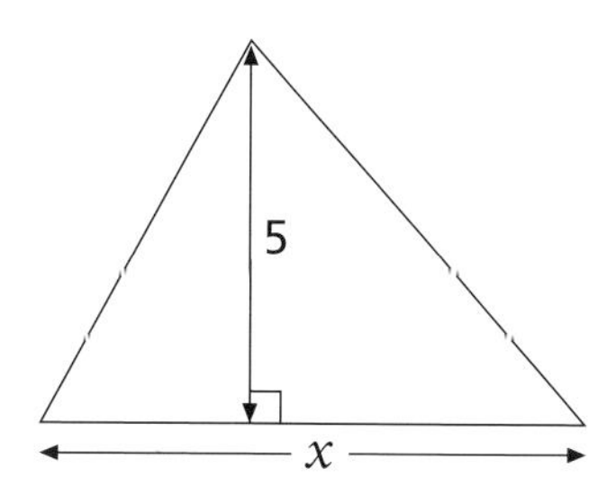

5 The triangle and rectangle both have the same area.
Find w.

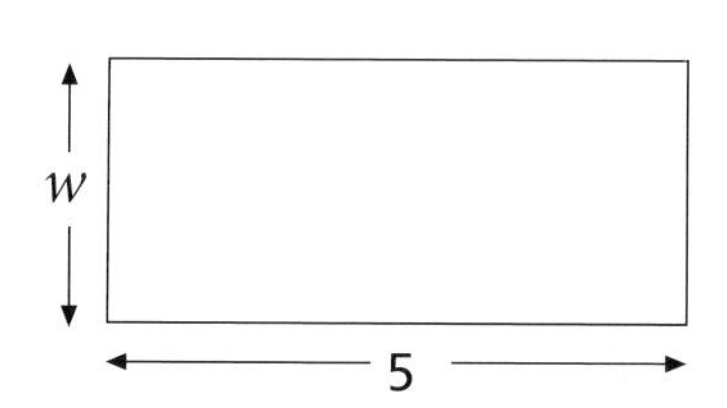

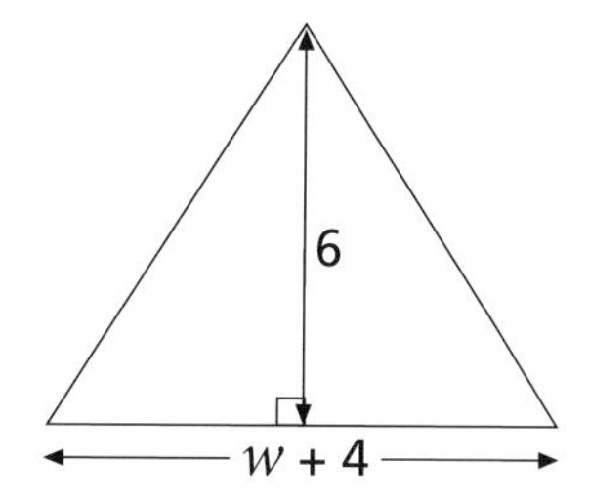

6 This shape has area $29\,\text{cm}^2$.
(It is not drawn to scale.)
Find d.

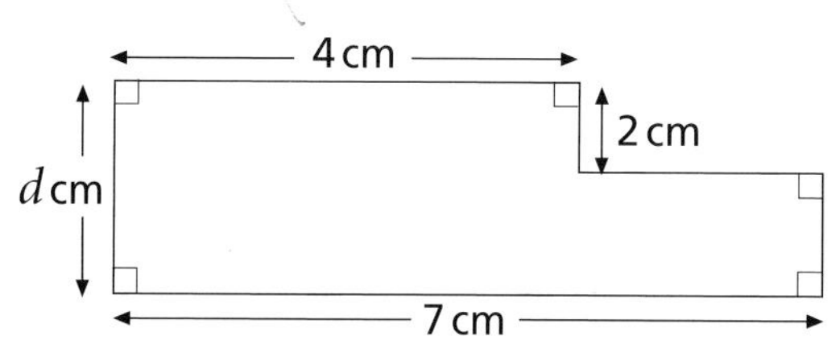

5 Solids

Section A

1 Each set of cross-sections below belongs to one of these tools, but they are in the wrong order.

For each set of cross-sections say what the tool is.

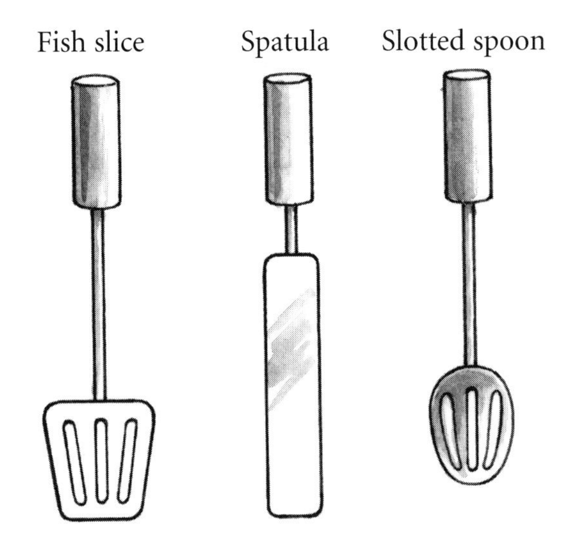

Set 1

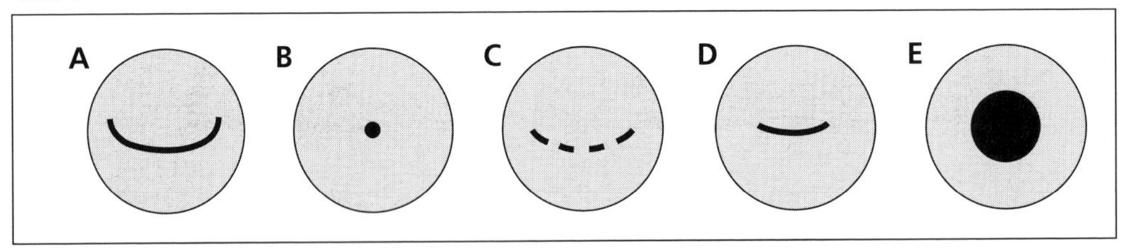

Set 2

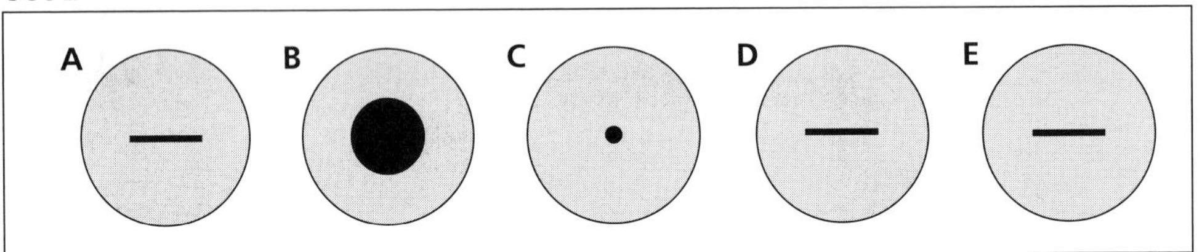

Set 3

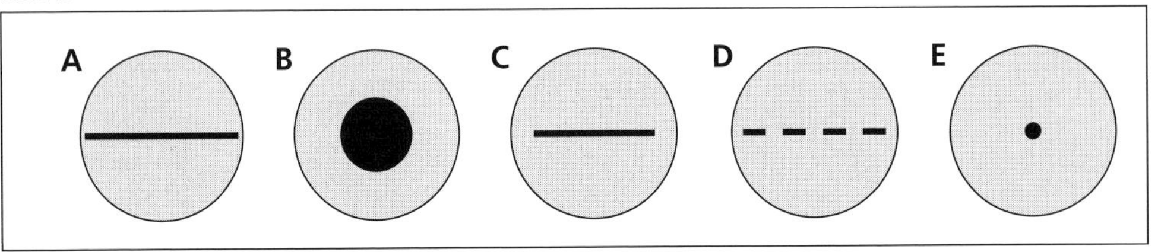

2 This key is held like this and lowered slowly into water.

Draw five or six cross-sections for the key as it is lowered.

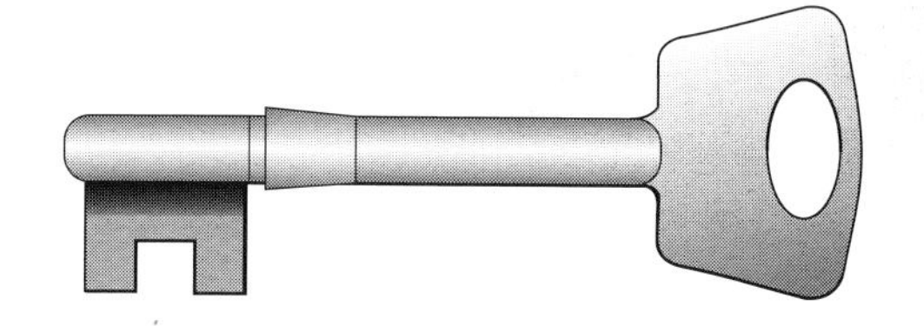

3 (a) Draw four cross-sections as this block is lowered into water like this.

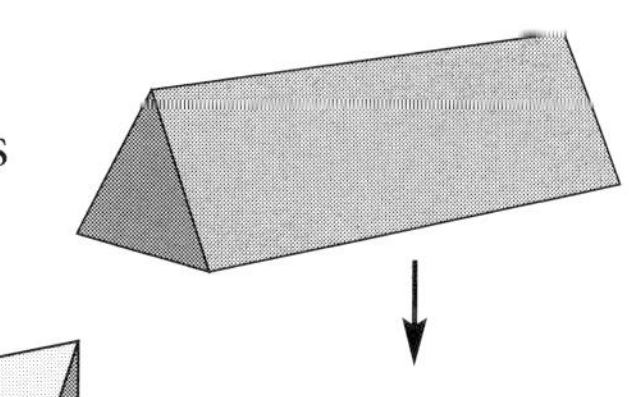

(b) Draw some cross-sections as it is lowered like this.

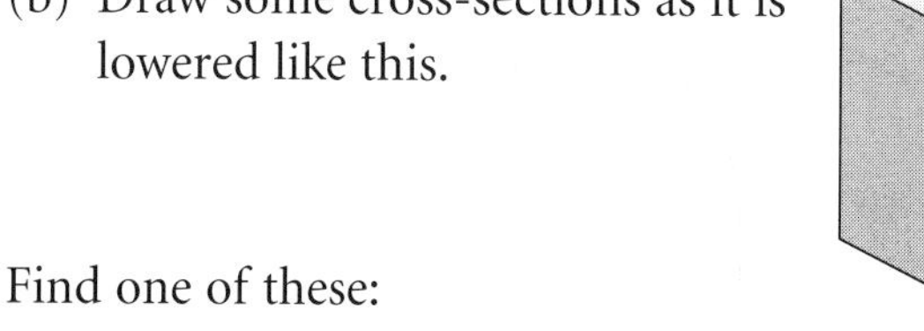

4 Find one of these:

a pencil a padlock a safety pin

(a) Imagine it is lowered into water. Draw a series of cross-sections.

(b) Imagine it is lowered a different way round. Draw a series of cross-sections.

5

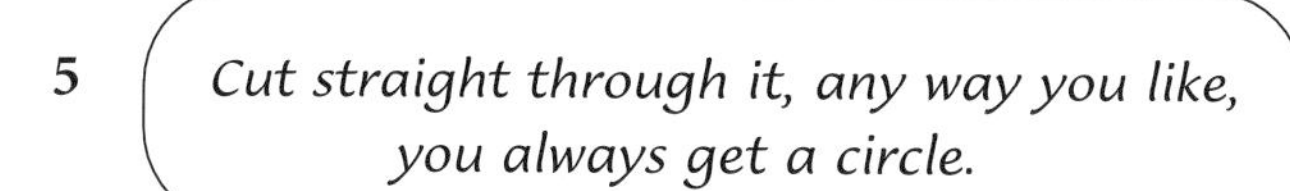

What solid shape is she talking about?

Section B

1 How many planes of symmetry do these solids have?

(a)

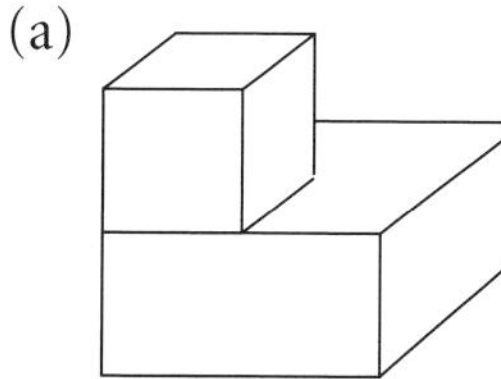

(b)

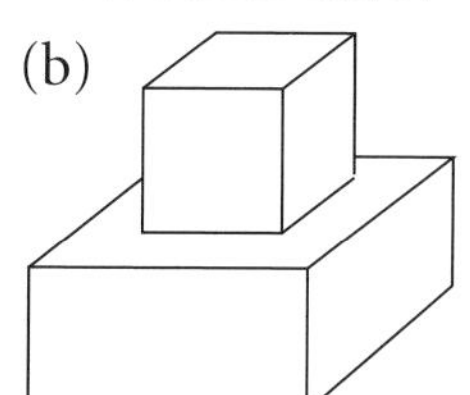

6 Units

Sections A and B

1 Calculate the area of each of these shapes.
Give each result in both m^2 and cm^2.

(a)

70 cm

1.9 m

(b)

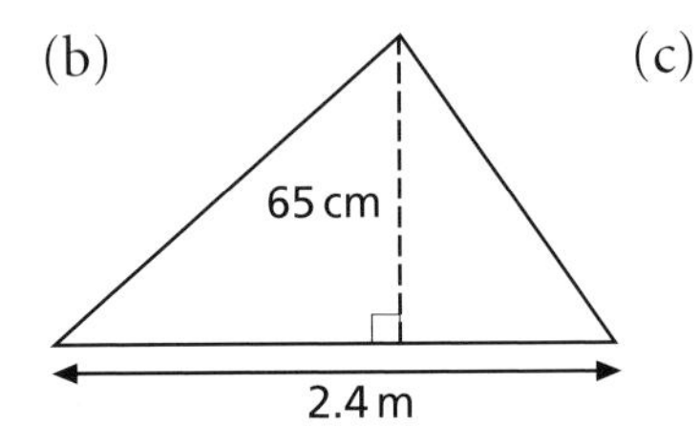

(c)

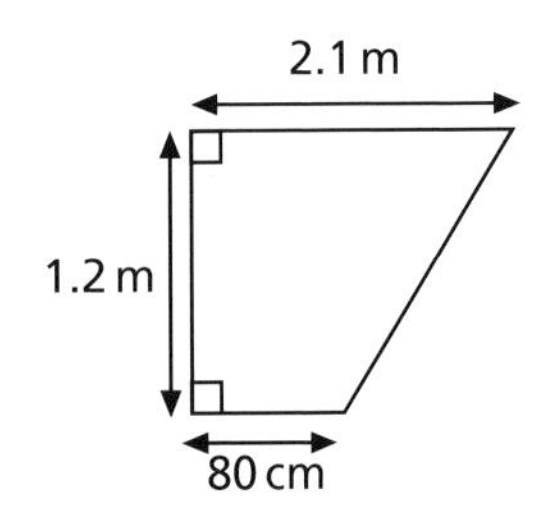

2 A rectangular tile measures 20 cm by 30 cm.
How many tiles are needed to cover an area of $4.5\,m^2$?

3 Write these quantities as litres.

(a) 100 ml (b) 5 ml (c) $1043\,cm^3$

(d) $650\,cm^3$ (e) $250\,cm^3$ (f) $5\,m^3$

4 How many litres would it take to fill each of these?

(a) four 500 ml yoghurt pots

(b) twenty 5 ml medicine spoons

(c) a petrol can 15 cm by 24 cm by 18 cm

(d) a fish tank 45 cm by 1.2 m to a depth of 40 cm

5 A swimming pool is 25 m long and 10 m wide.
The water is 1.5 m deep throughout.

(a) Find the volume of the pool in m^3.

(b) How many litres of water are needed to fill it?

(c) The pool can be filled at a rate of 800 litres per minute.
How long, to the nearest hour, will it take to fill the pool?

Sections C and D

1 Change these to kilometres.

(a) 40 miles (b) 105 miles (c) 255 miles

2 A swimming pool is 25 m long.
How many lengths should you swim to make a total distance of 5 km?

3 A petrol tank holds 40 litres.
Roughly how many gallons is this?

4 A sack of potatoes contains 20 kg.
Roughly how many pounds is this?

5 Jake uses 60 ml of washing liquid for each load of clothes.
How many loads can he wash using a 1.5 litre container of washing liquid?

6 Lynne says she is 1.4 m tall.
Rani says she is 5 feet tall.
Who is taller? Explain your answer.

7 The size of men's shirts is based on the collar size in inches.
Charles knows that his neck is 40 cm.
What size shirt should he buy?

7 Simplifying expressions

Section A

1 Which of these expressions is equivalent to $5p + 3 - 2p + 8$?

A $11 - 7p$

B $5 + 7p$

C $11 - 3p$

D $3p + 11$

2 Find four pairs of equivalent expressions.

J $8 - 7q - 3 - q$

K $6q - q + 8 - q - 3$

L $4q + 5$

M $8q - 11$

N $4q + 8 + 4q - 3$

O $5 - 8q$

P $2 + 10q - 2q + 3$

Q $5q - 4 - q + 4q - 7$

3 Simplify the following expressions.

(a) $4k + 5k - 3k$

(b) $8 - 2n + 4 + 3n$

(c) $12 - 7n - 5 + 6n$

(d) $6 + 4p - 3 + p - 2$

(e) $4n - 3 + 8 - 7n$

(f) $9r - 4r - 8 - 2r$

(g) $7n - 3n + 4 - 4n$

(h) $5t - 3 + 2t + 7 - 4t - 4$

(i) $4n - 5 + 3n - 2$

(j) $7 + 4v - 3v - 9$

4 Simplify the following expressions.

(a) $4u + 5v + 3u + 7v$

(b) $3x + 4y + 7y - 4x$

(c) $w - 3x - 4 + 3w - 5x$

(d) $2a + b - 3 + 4a - 4b$

(e) $3c + 7 - 2d + c - 2$

(f) $5e - 4f - 3e - 6 + 2f$

(g) $6g - 4 + 2g - 5h + 7$

(h) $k + m - 3m + 5 - 2k$

(i) $2n + 3 - 3p - 5 + 2n$

(j) $q + 4 - p + 7 - q - p$

Section B

1 Grid A

$6 - 2a$	$11 + 3a$	$2a + 4$
$5 + 5a$	$a + 7$	$9 - 3a$
10	$3 - a$	$4a + 8$

(a) (i) For grid A add the three expressions in each row, column and diagonal.

(ii) Explain why grid A will make a magic square for any value of a.

(b) From grid A make a magic square by

(i) using $a = 1$ (ii) using $a = 4$

(c) What value of a will give a magic square with 12 at the centre?

(d) Use grid A to find a magic square that has a total of 27.

2 (a) For each of grids B and C, add the three expressions in each row, column and diagonal.

(b) Which of these grids is a magic square?

Grid B

$2 + 3b$	$8b + 7$	$7b$
$10b + 1$	$3 + 6b$	$5 + 2b$
$5b + 6$	$4b$	$9b + 4$

Grid C

$9a + 2b$	$2a - 3b$	$10a - 2b$
$8a - 5b$	$7a - b$	$3b + 6a$
$4a$	$b + 12a$	$5a - 4b$

3 (a) Show that grid D is a magic square.

Grid D

$6 - d$	$4d + 5$	$3d + 10$
$11 + 6d$	$2d + 7$	$3 - 2d$
$d + 4$	9	$5d + 8$

(b) Use grid D to make a magic square

(i) with a magic total of 33

(ii) with 15 at the centre

4 For each of these, copy and complete the grid to make a magic square.

Grid E

		$7e$
$10e + 1$	$6e + 3$	$2e + 5$

Grid F

		$5f - g$
		$4f + 4g$
	$2f - 2g$	$9f + 3g$

Grid G

$4g - 1$		
$5g - 2$	$3g$	$g + 2$

Grid H

$6h - 4j$		
	$5h - j$	$3h - 3j$
		$4h + 2j$

Section C

1 Copy and complete each of these walls.

(a)

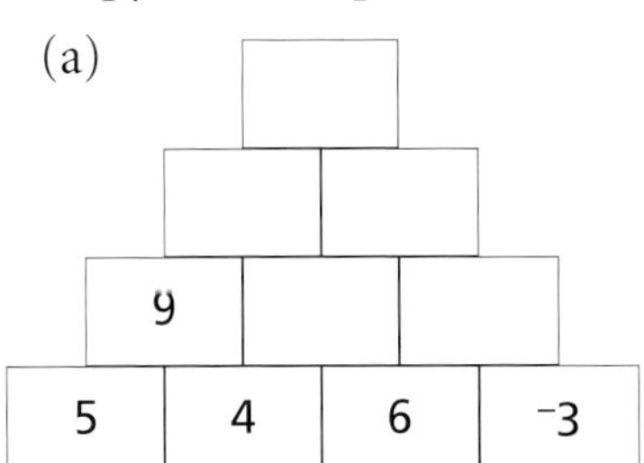

(b)

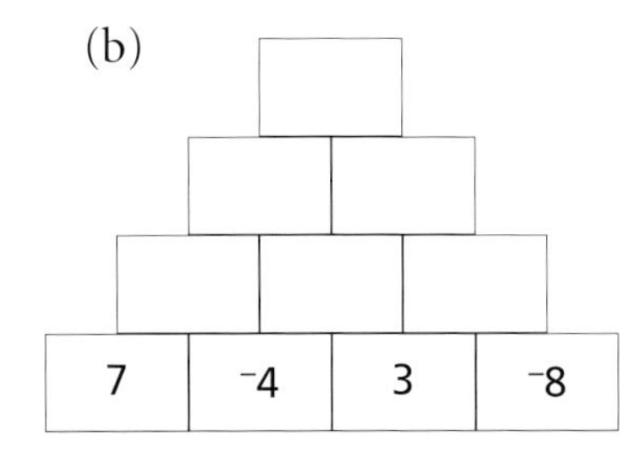

2 Work out the value of the letter in each of these.

(a)

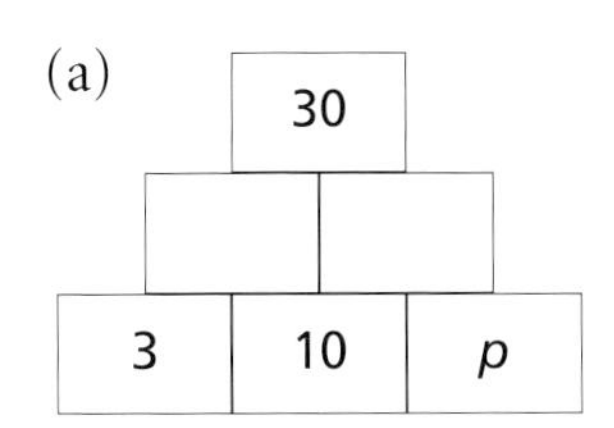

(b)

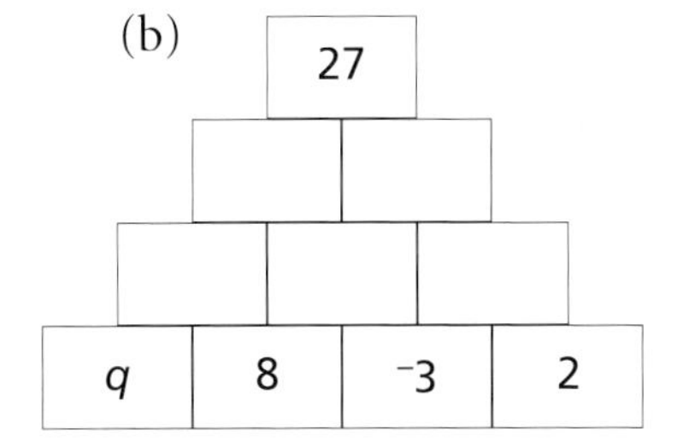

(c)

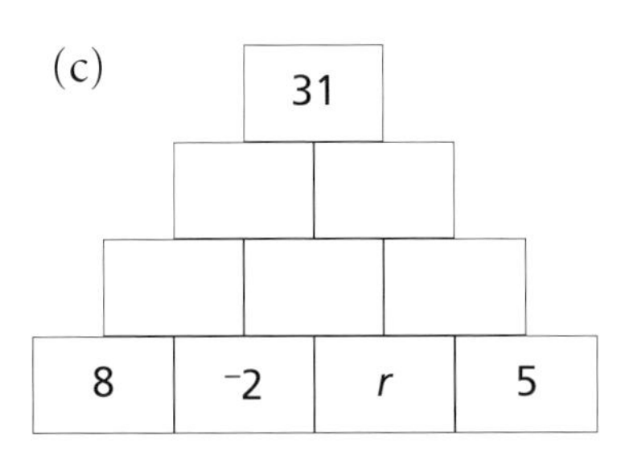

(d)

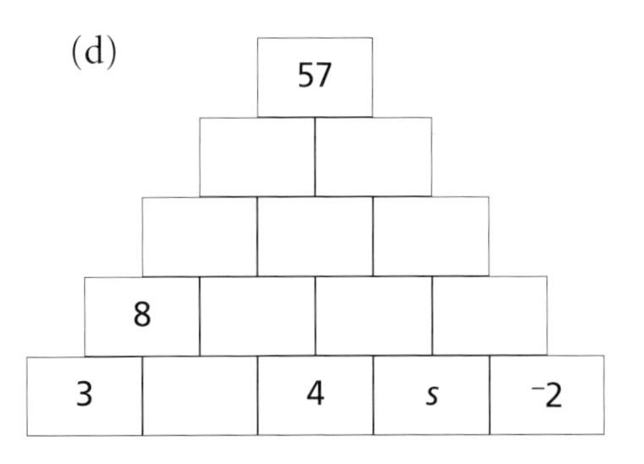

3 Find all the missing numbers in each of these walls.

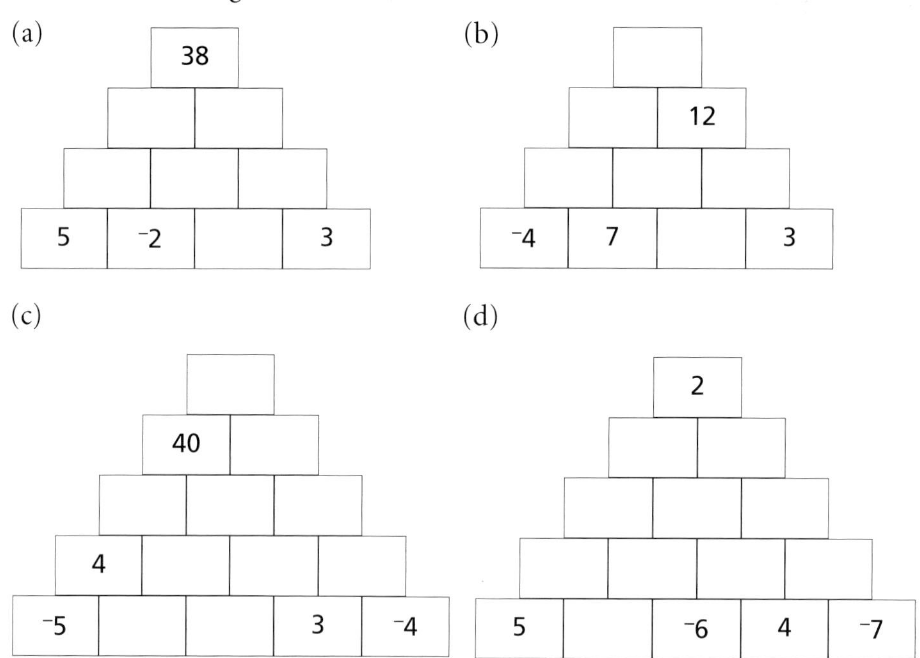

4 In this wall the missing numbers on the bottom row are the same.
What are the numbers on the bottom row ?

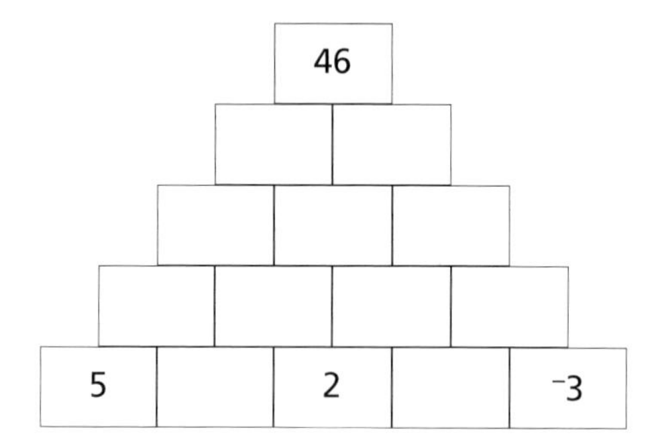

8 Fractions and decimals

Sections A and B

1 Pair off equivalent fractions in each of these lists.

(a) $\frac{6}{9}$ $\frac{2}{4}$ $\frac{8}{10}$ $\frac{5}{10}$ $\frac{12}{15}$ $\frac{8}{12}$

(b) $\frac{5}{15}$ $\frac{18}{24}$ $\frac{10}{12}$ $\frac{6}{8}$ $\frac{5}{6}$ $\frac{2}{6}$

(c) $\frac{15}{40}$ $\frac{8}{32}$ $\frac{3}{12}$ $\frac{9}{24}$ $\frac{2}{6}$ $\frac{4}{12}$

(d) $\frac{12}{30}$ $\frac{1}{6}$ $\frac{3}{12}$ $\frac{4}{24}$ $\frac{5}{20}$ $\frac{6}{15}$

2 Copy and complete these.

(a) $\frac{240}{360} = \frac{?}{72} = \frac{6}{?} = \frac{?}{3}$ (b) $\frac{360}{480} = \frac{?}{120} = \frac{30}{?} = \frac{?}{4}$

3 Which fraction in each pair is greater?

(a) $\frac{1}{4}$, $\frac{3}{20}$ (b) $\frac{2}{3}$, $\frac{15}{21}$ (c) $\frac{7}{8}$, $\frac{9}{10}$ (d) $\frac{5}{12}$, $\frac{2}{5}$

4 Work these out. Simplify the results where possible.

(a) $\frac{1}{4} + \frac{2}{3}$ (b) $\frac{3}{4} - \frac{1}{5}$ (c) $\frac{2}{5} + \frac{1}{3}$ (d) $\frac{3}{10} + \frac{1}{6}$

(e) $\frac{4}{9} - \frac{1}{6}$ (f) $1\frac{1}{2} + \frac{4}{9}$ (g) $2\frac{11}{16} - 1\frac{3}{4}$ (h) $3\frac{5}{8} - 2\frac{5}{6}$

(i) $4\frac{5}{8} + 1\frac{2}{3}$ (j) $6\frac{11}{16} + 1\frac{9}{10}$

5 Work these out but do not simplify your results.

$\frac{1}{2} + \frac{1}{4}$ $\frac{1}{2} + \frac{1}{4} + \frac{1}{8}$ $\frac{1}{2} + \frac{1}{4} + \frac{1}{8} + \frac{1}{16}$

The three results form a sequence.
What are the next two terms in the sequence?

6 Repeat question 5 for each of these.

(a) $\frac{1}{2} - \frac{1}{4}$ $\frac{1}{2} - \frac{1}{4} + \frac{1}{8}$ $\frac{1}{2} - \frac{1}{4} + \frac{1}{8} - \frac{1}{16}$

(b) $\frac{1}{3} + \frac{1}{6}$ $\frac{1}{3} + \frac{1}{6} + \frac{1}{12}$ $\frac{1}{3} + \frac{1}{6} + \frac{1}{12} + \frac{1}{24}$

(c) $\frac{1}{3} - \frac{1}{6}$ $\frac{1}{3} - \frac{1}{6} + \frac{1}{12}$ $\frac{1}{3} - \frac{1}{6} + \frac{1}{12} - \frac{1}{24}$

Sections C and D

1 Work these out.

(a) $4 \times \frac{1}{2}$ (b) $9 \times \frac{1}{4}$ (c) $11 \times \frac{1}{2}$ (d) $12 \times \frac{1}{5}$

(e) $5 \times \frac{1}{3}$ (f) $8 \times \frac{2}{3}$ (g) $\frac{4}{5} \times 7$ (h) $6 \times \frac{3}{4}$

(i) $\frac{5}{6} \times 18$ (j) $4 \times \frac{9}{10}$ (k) $\frac{3}{7} \times 4$ (l) $9 \times \frac{5}{8}$

2 How many

(a) $\frac{1}{3}$s in 5

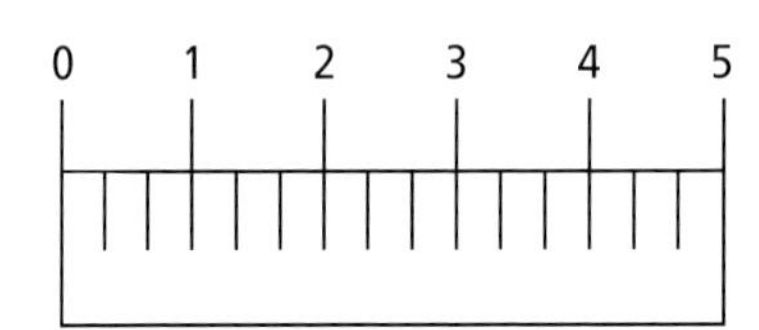

(b) $\frac{2}{5}$s in 4

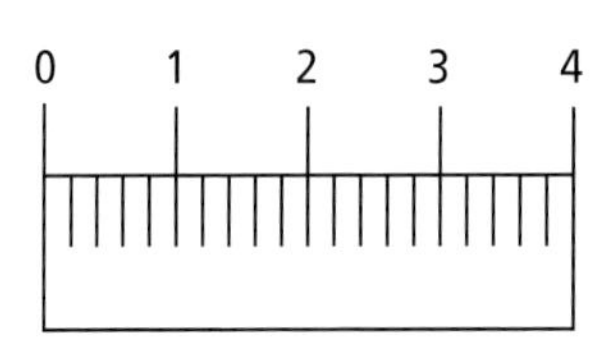

Write down the value of each of these.

(c) $5 \div \frac{1}{3}$ (b) $4 \div \frac{2}{5}$

3 Work these out.

(a) $3 \div \frac{1}{4}$ (b) $6 \div \frac{1}{2}$ (c) $5 \div \frac{1}{3}$ (d) $10 \div \frac{1}{5}$

4 Work these out.

(a) $9 \div \frac{2}{3}$ (b) $6 \div \frac{3}{4}$ (c) $5 \div \frac{5}{8}$ (d) $12 \div \frac{2}{5}$

5 Work out $3 \div \frac{2}{3}$.
Draw a diagram to show that the result is correct.

6 Work these out.

(a) $9 \times \frac{3}{4}$ (b) $9 \div \frac{3}{4}$ (c) $10 \div \frac{5}{6}$ (d) $\frac{5}{6} \times 10$

Sections E and F

1 (a) Without using a calculator, work out $\frac{1}{15}$ as a recurring decimal. Show all your working clearly.

(b) Use the result for $\frac{1}{15}$ to find the recurring decimal for $\frac{1}{30}$.

2 Change each of these fractions to a decimal.

(a) $\frac{7}{25}$ (b) $\frac{5}{6}$ (c) $\frac{3}{8}$ (d) $\frac{7}{9}$

(e) $\frac{1}{40}$ (f) $\frac{3}{11}$

3 Sam spends $\frac{1}{3}$ of the day sleeping, $\frac{1}{2}$ of the day working, and has the rest free.

(a) What fraction of Sam's day is free?

(b) How long does he have free?

4 On 4 days of the week, Mary cycles for $\frac{3}{4}$ hour.
On the other 3 days, she runs for $\frac{3}{4}$ hour and swims for $\frac{1}{2}$ hour.

(a) How many hours exercise does she do each week?

(b) How much more exercise will she do if she cycles on 3 days, and runs and swims on 4 days?

5 Work each of these out (i) as a fraction
(ii) as a recurring decimal

(a) $\frac{1}{6} + \frac{1}{2}$ (b) $\frac{1}{3} + \frac{1}{5}$

Mixed questions 1

1 This is the plan of a rectangular field, drawn to a scale of 1 cm to 50 m.

Calculate

(a) the perimeter of the field in kilometres

(b) the area of the field in km^2

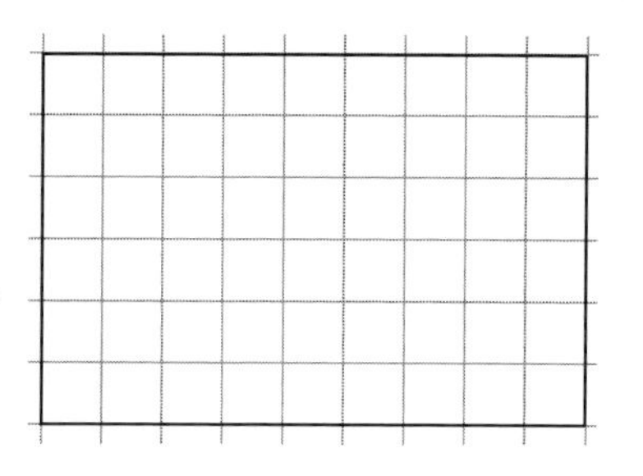

2 Solve these equations.

(a) $x + 12 = 5x + 10$ (b) $4y - 13 = 7y + 2$ (c) $15z + 37 = 7z + 93$

(d) $3(a + 4) = 5(a - 2)$ (e) $\frac{1}{2}(8 + b) = b - 6$ (f) $3(9 - c) = 2(c + 1)$

3 (a) Show by dividing that $\frac{1}{6}$ as a recurring decimal is 0.1666…

(b) Hence show that $\frac{1}{6} - \frac{1}{10} = 0.0666\ldots$

(c) Work out $\frac{1}{6} - \frac{1}{10}$ as a fraction, in its simplest form.

4 Simplify these expressions.

(a) $5x + 2 - x - 9$ (b) $4y - 3 - 7b - 5$ (c) $14 - 5z - 9 + 2z$

5 Sketch the cross-section of this object at these heights above its base.

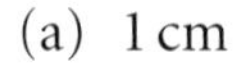

(a) 1 cm

(b) 2 cm

(c) 3 cm

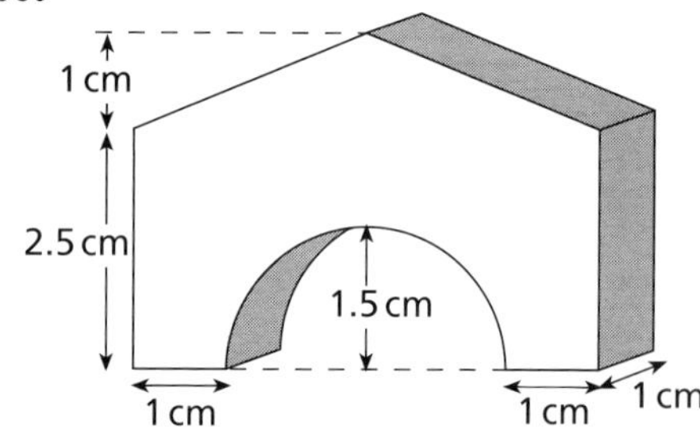

6 This is an enlargement of the end of a pencil lead.
The diameter of the real lead is 0.5 mm.

What is the scale factor of the enlargement?

9 Transformations

Sections A and B

1 Copy these diagrams on to squared paper and draw the reflections in the mirror lines.

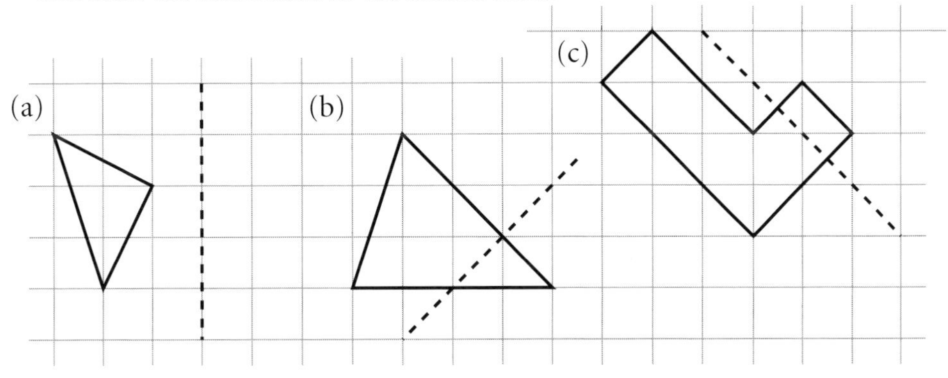

2 Copy these diagrams on to squared paper and draw in the mirror lines.

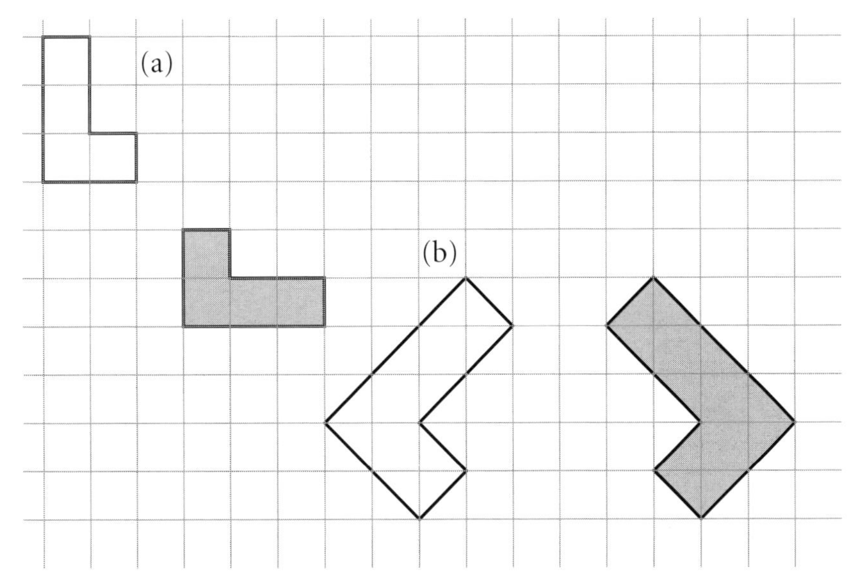

3 What are the equations of the mirror lines for each of these reflections?

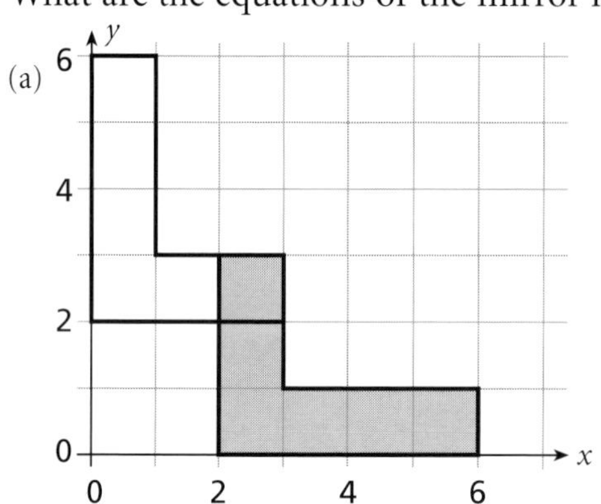

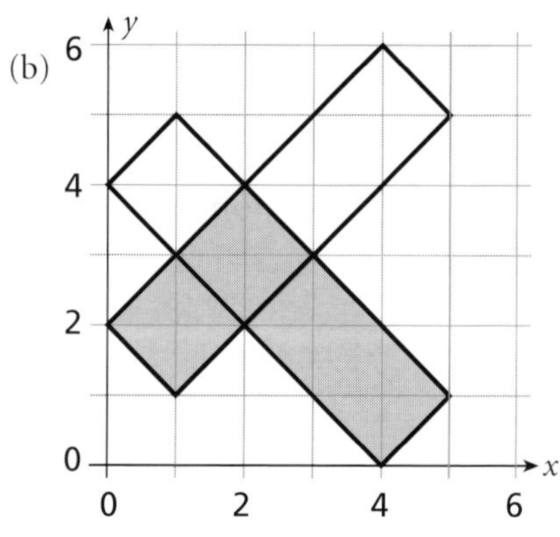

Sections C and D

1 Copy these diagrams on to squared paper and rotate each shape about the given point, using the angle given.

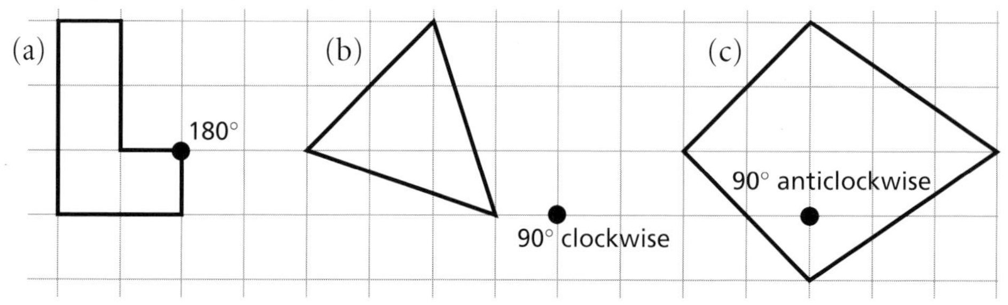

2 Copy these diagrams on to squared paper and mark the centre of rotation. Describe each rotation fully. (The image is shaded.)

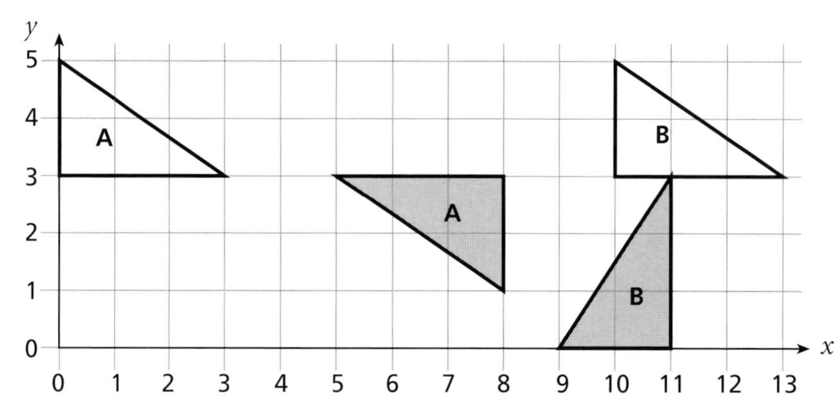

3 (a) What are the coordinates of the centre of this 180° rotation of a flag?

(b) Where would the centre of a 180° rotation be if the image of point P was at (3, 4)?

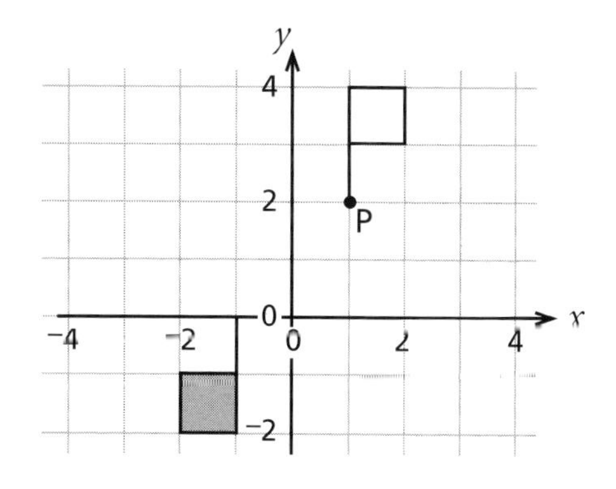

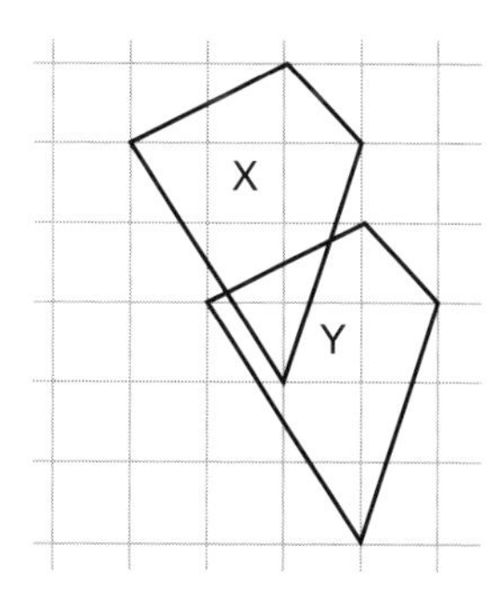

4 What transformation moves

(a) shape X on to shape Y

(b) shape Y on to shape X

Section E

1 Copy this diagram.
Draw the enlargement of triangle JKL with centre C and scale factor 2.

Write down the coordinates of the vertices of the enlargement.

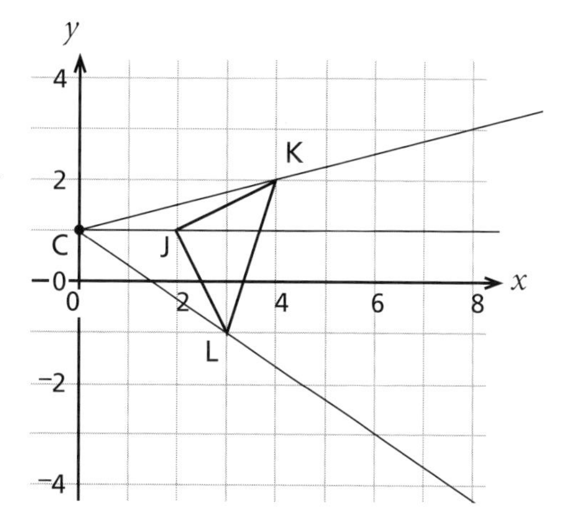

2 Using a new set of axes, draw the enlargement of triangle JKL with centre (3, 1) and scale factor 3.

Write down the coordinates of the vertices of the enlargement.

Section F

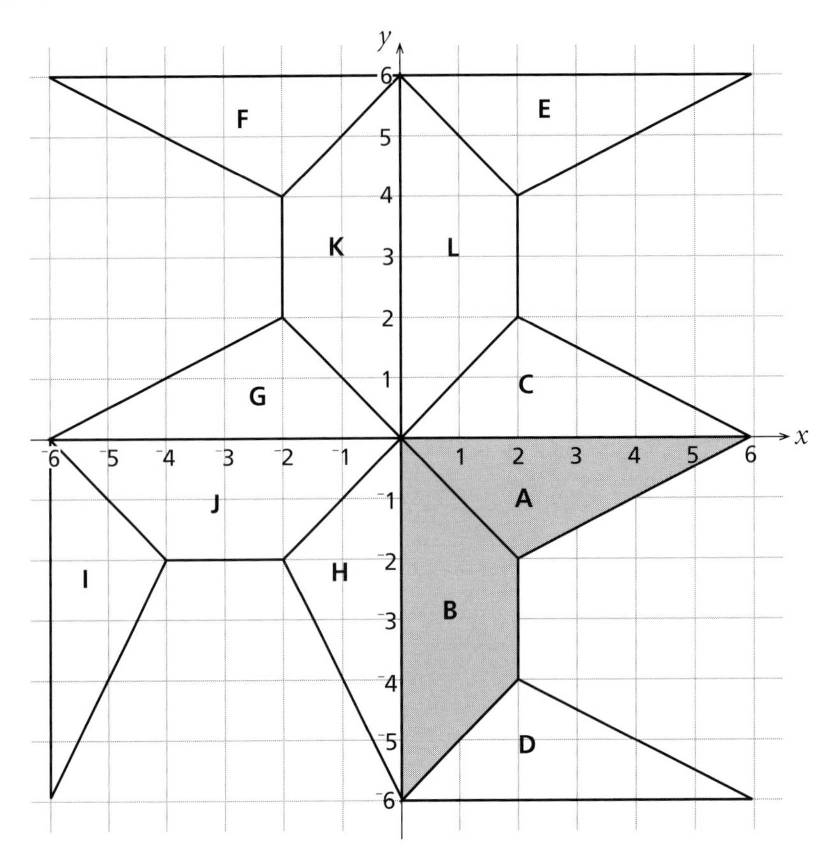

1 The above design is made up of transformations of a triangle A and a trapezium B.

Describe fully each of the following transformations.

(a) A to C (b) A to D (c) A to E

(d) C to F (e) A to G (f) A to H

(g) B to J (h) B to K (i) B to L

(j) C to H (k) G to F (l) G to H

(m) D to F (n) J to L

Section H

1 (a) Copy this diagram.

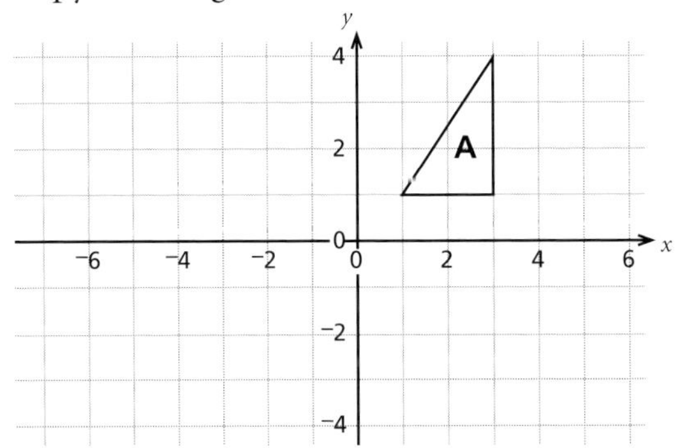

Draw the image of A after reflection in $x = 4$ then reflection in the y-axis.

Label the image B.

(b) Describe the single transformation which has the same effect as these two.

2 (a) Copy this diagram.

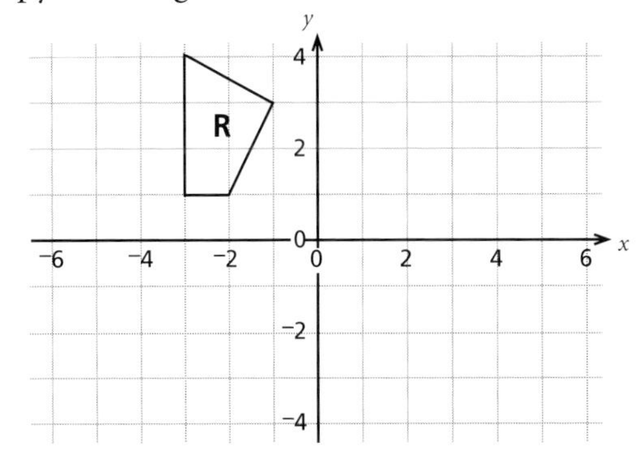

Draw the image of R after a rotation of 180° about (1, 1) then a rotation of 180° about (0, ⁻1).

Label the image S.

(b) Describe the single transformation which has the same effect as these two.

Linear equations and graphs

Section A

1 Find the gradient of each of these lines.

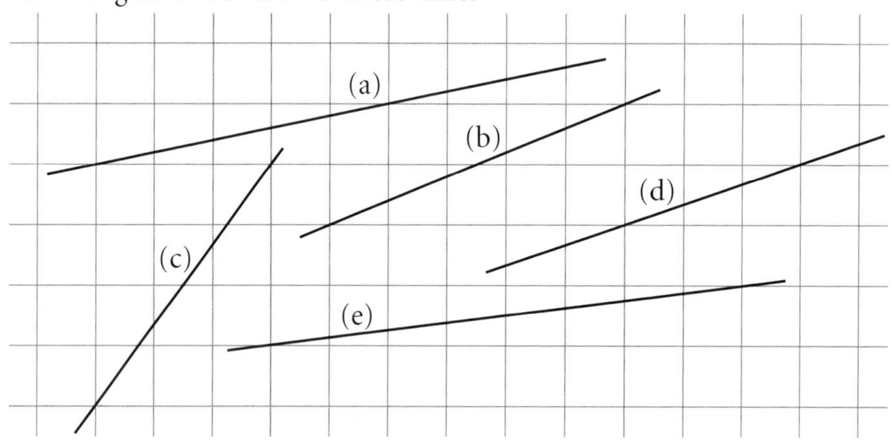

2 The desirable maximum gradients for different types of road are as follows:

motorway	0.03
dual carriageway	0.04
single carriageway	0.06

Find the gradient of each of these roads to two decimal places.
Write down a type of road which would be suitable for each one.

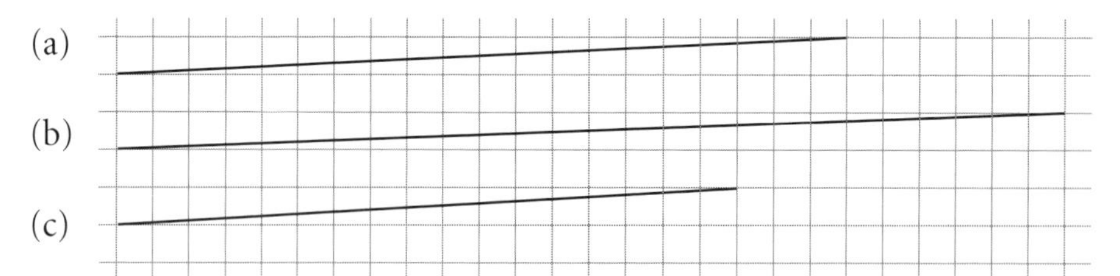

Section C

1 Write down (i) the gradient (ii) the intercept on the y-axis of each of these linear graphs.

(a) $y = 4x + 3$ (b) $y = \frac{x}{2}$ (c) $y = 3x - 1$

(d) $y = 4$ (e) $y = 0.1x - 3$ (f) $y = 7x + 6$

2 Write down the equation of each of these linear graphs.

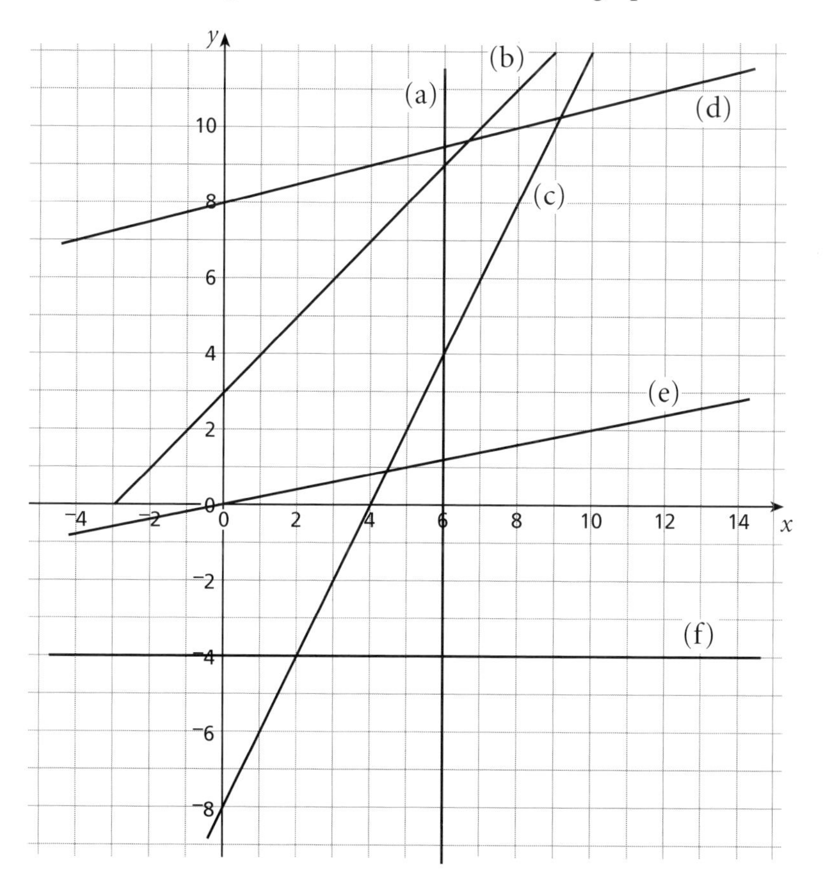

Section D

1 Write down (i) the gradient (ii) the intercept on the y-axis of each of these linear graphs.

(a) $y = {}^{-}3x$ (b) $y = 4 - x$ (c) $y = 5 + 2x$

(d) $y = {}^{-}0.2x - 3$ (e) $y = 3 - 4x$ (f) $y = 2x - \frac{1}{2}$

2 This sketch shows four graphs, A, B C, and D.
Below are four equations.
Match each graph with the correct equation.

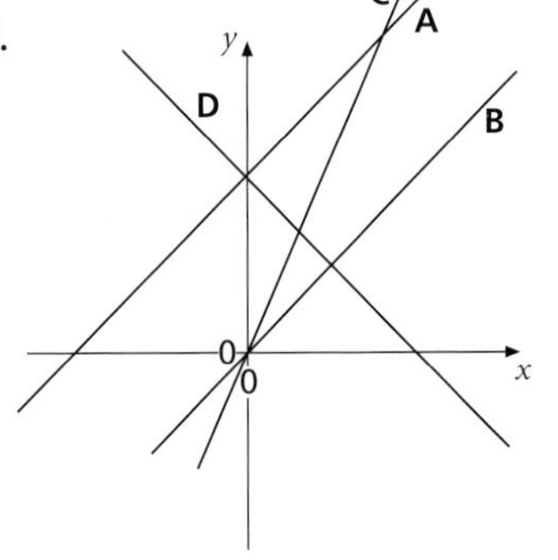

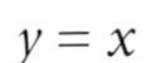

$y = x$ $y = 2x$

$y = x + 2$ $y = 2 - x$

3 This diagram shows a sketch of the graph of $y = 3x - 2$.
Copy it about twice this size.

On your copy also sketch graphs of the lines $y = 4x - 2$, $y = x - 2$ and $y = {}^{-}3x - 2$.

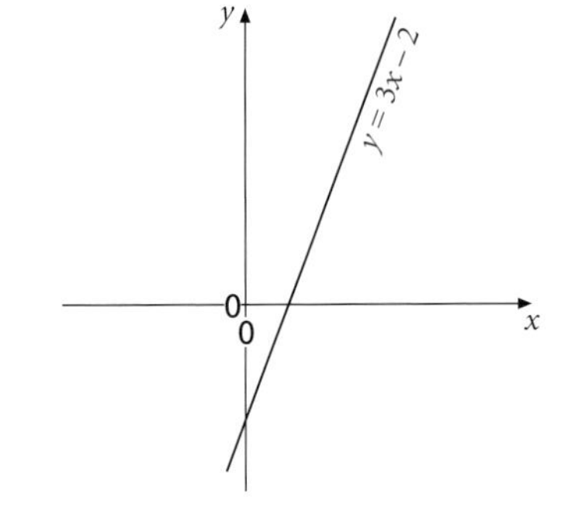

4 This diagram shows a sketch of the graph of $y = {}^{-}2x + 3$.
Copy it about twice this size.

On your copy also sketch graphs of the lines $y = {}^{-}2x + 5$, $y = 2x$ and $y = {}^{-}x + 3$.

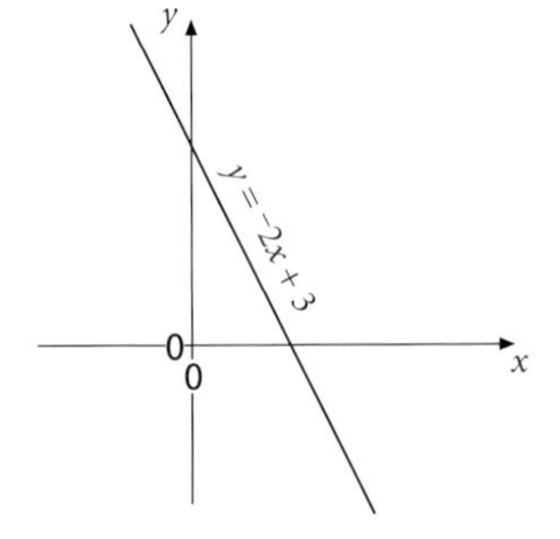

12 Percentage changes

Section A

1 Write each of these percentages as a decimal.

(a) 55% (b) 60% (c) 65.5% (d) 84.6%

2 Write each of these decimals as a percentage.

(a) 0.39 (b) 0.4 (c) 0.425 (d) 0.509

3 Write each of these percentages as a decimal.

(a) 4% (b) 4.2% (c) 6.5% (d) 0.8%

4 Write each of these decimals as a percentage.

(a) 0.03 (b) 0.035 (c) 0.048 (d) 0.1825

5 Match up these percentages, decimals and test scores.
You should end up with three lists (not all the same length).

8% 80 out of 100 0.8% 8 out of 10 0.08
8 out of 100 80% 0.8 0.008 8 out of 1000

6 Calculate each of these to the nearest penny.

(a) 28% of £120 (b) 12.5% of £150 (c) 43.9% of £80
(d) 6.5% of £48 (e) 3.4% of £95 (f) 0.7% of £42

7 Change $\frac{9}{40}$ to (a) a decimal (b) a percentage

8 Jody's mark in a test was 27 out of 40. What percentage is this?

9 Anna won 16 of her last 23 tennis matches. What percentage is this?
Give your answer to the nearest 0.1%.

10 Match up these percentages, decimals and test scores.
You should end up with two lists.

1.5% 15 out of 100 15% 3 out of 200 12 out of 80
3 out of 20 0.015 $1\frac{1}{2}$ out of 100 0.15 15 out of 1000

Sections B and C

1 Membership fees for adults are to go **up** by 12%.

Calculate the new fee for adults if the present fee is

(a) £25 (b) £40 (c) £18 (d) £110

2 Membership fees for children are to go **down** by 16%.

Calculate the new fee for children if the present fee is

(a) £5 (b) £2.50 (c) £35 (d) £14.50

3 If prices are **increased** by 12.5%, what number do you multiply by to get the new price?

4 The prices of cinema tickets are increased by 12.5%.

Calculate the new price if the old one is

(a) £4 (b) £2.40 (c) £12 (d) £3.60

5 Prices are **reduced** by 7.5%.
What number would you multiply by to find the reduced price?

6 In a sale, prices are reduced by 7.5%.

Calculate the new price if the old one is

(a) £40 (b) £8.80 (c) £16 (d) £360

7 What number do you multiply by, to increase prices by

(a) 17% (b) 26.5% (c) 7% (d) 3.6%

8 What number do you multiply by, to reduce prices by

(a) 18% (b) 3% (c) 10.5% (d) 4.8%

9 For each of these multipliers state whether they give an increase or a decrease and give the percentage change.

(a) 1.35 (b) 1.6 (c) 0.75 (d) 0.91

(e) 0.3 (f) 1.165 (g) 0.955 (h) 1.025

Section D

1 Mark was paid £8 per hour then his rate was increased by 5%.
Six months later he had another rise of 10%. What was his rate of pay

(a) after the first increase (b) after the second increase

2 Prices were reduced by 10% in a sale. The sale prices were reduced by 15% for clearance.

Find the clearance price of items that before the sale cost

(a) £180 (b) £350 (c) £72 (d) £12

3 The prices of motorbikes went up by 18% in May and then were later reduced by 12% in a sale.

What would the sale price be for motorbikes that originally cost

(a) £4800 (b) £12 600 (c) £6250 (d) £25 000

4 Calculate the final prices, rounding your answers to the nearest penny if necessary.

(a) £250 increased by 9% then increased by 1%

(b) £45 increased by 12.5% then decreased by 10%

(c) £1800 decreased by 4.2% then increased by 1.5%

(d) £56 000 decreased by 8% then decreased by 6%

5 (a) Musicfayre increased its prices by 10% in April and then decreased them by 10% in July.
Will prices now be the same as before April, or higher, or lower?
Explain.

(b) SoundOn decreased its prices by 10% in April and then increased them by 10% in July.
Before April a CD player cost the same in both shops.
Will it cost the same in both shops after July? Explain.

Section E

1 The price of a pair of trousers goes up from £25 to £28.

(a) What is the multiplier that changes 25 to 28?

(b) What is the percentage increase?

2 The number of members of a youth club dropped from 75 to 69.

(a) What is the multiplier that changes 75 to 69?

(b) What is the percentage decrease?

3 Calculate the percentage change for each of these prices and say whether it was an increase or a decrease.

Give your answer to the nearest 1% where necessary.

(a) £15 to £15.60

(b) £3.50 to £3.92

(c) £44 to £42.24

(d) £7200 to £5256

(e) £32.90 to £40.80

(f) £18 500 to £16 800

4 Calculate each of these percentage changes, stating whether it is an increase or a decrease and giving your answers to the nearest 0.1% where necessary.

(a) £12 000 to £10 500

(b) £7.20 to £7.45

(c) £450 to £482.50

(d) £15.99 to £12.50

5 The population of tortoises on a tropical island was 450 a year ago and is 415 now.

(a) What is the percentage decrease in the population of tortoises (to the nearest 0.1%)?

(b) If this rate of decrease continues, what will the tortoise population be next year? And the year after?
(Give your answers to the nearest whole tortoise!)

Section F

1 Tony's earnings increase from £34 to £36.90.
John's earnings increase from £21.50 to £23.44.

Does Tony or John get the greater percentage increase?
Justify your answer.

2 Laura invests £50 in a savings account.
During each year the amount in the account goes up by 4.5%.
If she does not take any of the money out of the account,
how much will it be worth after

(a) one year (b) five years

3 The population of the world is approximately 6 billion and it is estimated that it will increase by 78 million next year.

(a) What will the world's population be next year?

(b) What is the percentage rate of growth for the population of the world?

4 The number of cases of a particular disease has been going down at the rate of 25% each year.
Karen says that in two years time the occurrence of the disease will have halved.

Is she correct? Explain.

5 The population of wild birds that visit Andrew's garden has been decreasing at the rate of 20% each year.
Andrew estimates that in two years' time the population will have gone down another 40%.

Will his estimate put the population too high, too low or about right?
Explain.

6 If a population is decreasing at 10% per year, approximately how many years will it take for the population to halve?

14 Probability from experiments

Section B

1 Maureen works in a dress factory.
She has to reject any dress that is not perfect.
One day she checks 200 dresses and rejects 30 of them.
Estimate the probability that a dress is perfect.

2 Gary records the numbers of times Starbank United win, lose and draw in a season. These are his results.

W W D W L W D D L W L L D W L
L D L L W D D W W L W D D L W

Estimate the probability that Starbank United will

(a) win (b) lose (c) draw

3 Zara and Naomi enjoy playing chess together.
They record who wins each game. Here are the results.

Z N N Z Z Z Z N Z N N Z Z N N N Z N Z Z
N N Z N Z Z N N N N N N N Z Z N Z Z Z N

What is the relative frequency of each of these?

(a) Naomi winning (b) Zara winning

Who seems to be the better chess player?

4 Jasmine runs the school tuck shop.
She records the flavours of crisps that are sold one break time.
Here is her record at the end of break.

Flavour	Tally
Ready salted	𝍸 𝍸 II
Salt and vinegar	𝍸 𝍸 𝍸 III
Cheese and onion	𝍸 𝍸 𝍸 𝍸 I
Prawn cocktail	𝍸 𝍸 𝍸 𝍸 III
Beef	𝍸 II
Pickled onion	𝍸 𝍸

Work out the relative frequency for each flavour.
Write them as decimals, to two decimal places.

Section E

1 A fair spinner is coloured like this.

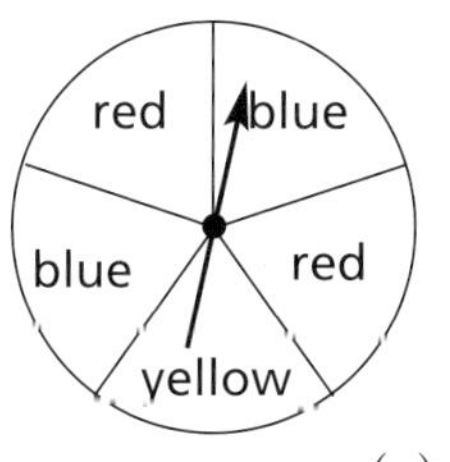

If the arrow is spun 150 times, roughly how many times would you expect it to show

(a) red (b) yellow (c) blue

2 An ordinary dice is thrown 500 times.
Roughly how many times would you expect it to show

(a) 5 (b) an odd number (c) a multiple of 3

(d) a square number (e) a factor of 12

3 Chris travels to work on the bus for 70 days.
He arrives late on 23 of the days.

(a) Estimate the probability that Chris is late for work.

(b) If Chris travels to work on the bus for 200 days, roughly how many times should he expect to be late?

4 Over the past few days, a shop sold the following numbers of packets of crisps.

Ready salted	83
Salt and vinegar	52
Cheese and onion	41
Prawn cocktail	27
Beef	19
Pickled onion	13

(a) Estimate the probability that the next packet to be sold is ready salted.

(b) The shop is going to order 1000 packets of crisps.
About how many of each flavour should they order?

15 Bearings

Section A

1 Draw the rectangle ABCD to a scale of 1 cm to 4 km.

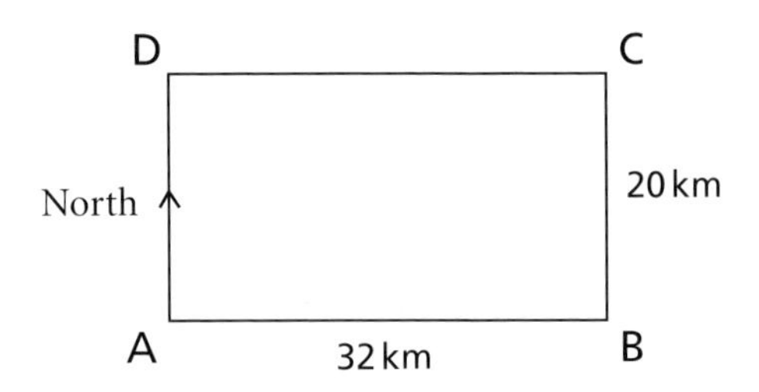

(a) What is the bearing of B from A?

(b) What is the bearing of C from A?

(c) What is the bearing of D from A?

(d) Find and label the point K whose bearing from A is 065° and from B is 325°.

(e) What is the bearing of K from D?

(f) How far is it from K to D?

2 This sketch map shows the route of a sailing race. It is in the shape of an equilateral triangle.

AB is on a bearing of 070°.

(a) Calculate the bearing of BC.

(b) Calculate the bearing of CA.

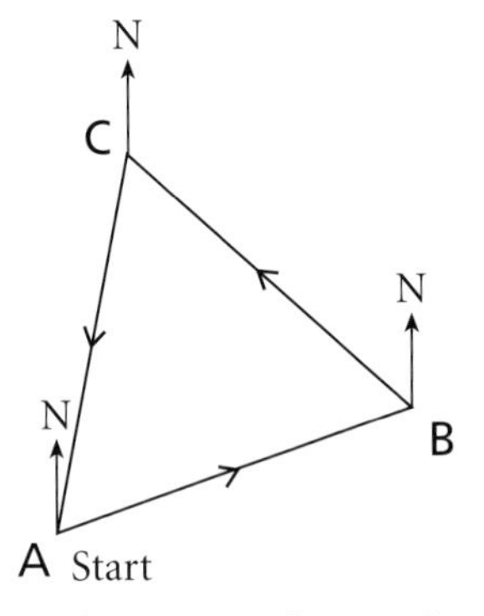

3 This diagram shows a square ABCD. B is on a bearing of 060° from A.

What is the bearing of

(a) C from A

(b) D from A

(c) B from E

(d) A from E

(e) D from C

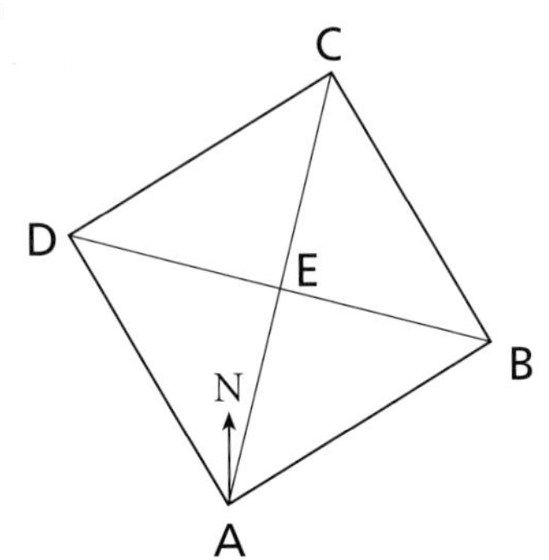

16 Forming equations

Section A

1 Alan, Ben, Chloe and Daniel have some sweets.

Suppose Alan has x sweets.

(a) Write down expressions for the number of sweets that

(i) Ben has (ii) Chloe has (iii) Daniel has

(b) Tina says that an expression for her number of sweets is $3x + 1$.
Write a sentence in words for Tina's sweets.

2 Each of these bags has n sweets in it.
1 sweet is taken out of each bag.

Write an expression for the total number of sweets left.

3 A school minibus carries 12 people.

(a) Suppose there are s minibuses full of people.
Write an expression for the total number of people.

(b) Suppose there are t minibuses all full except for one which has only 4 passengers.
Write an expression for the total number of people.

Section B

1 Iain and William breed pet mice.
Suppose Iain has q mice.
William has 6 mice fewer than Iain.

(a) Write an expression for the number of mice William has.

(b) Altogether Iain and William have 84 mice.
Write an equation for the total number of mice.

(c) Solve the equation.
How many mice does Iain have?
How many mice does William have?

(d) Check that your answers work.

2 This rectangle has lengths of sides as shown.

$(x + 5)$ cm

x cm

(a) Write down an expression for the perimeter of the rectangle.
Make it as simple as possible.

(b) The perimeter of the rectangle is 74 cm.
Form an equation.

(c) Solve the equation.
What are the lengths of the sides of the rectangle?

3 Imran, Rehana and Roisin have some bulbs to plant.
Rehana has half as many bulbs as Imran.
Roisin has 100 more bulbs than Imran.
Suppose Imran has n bulbs.

(a) Write an expression for the number of bulbs Rehana has.

(b) Write an expression for the number of bulbs Roisin has.

(c) Altogether they have 1000 bulbs to plant.
Form an equation and solve it to find n.

(d) Write down how many bulbs each person has.

4 The angles of this triangle are as shown.

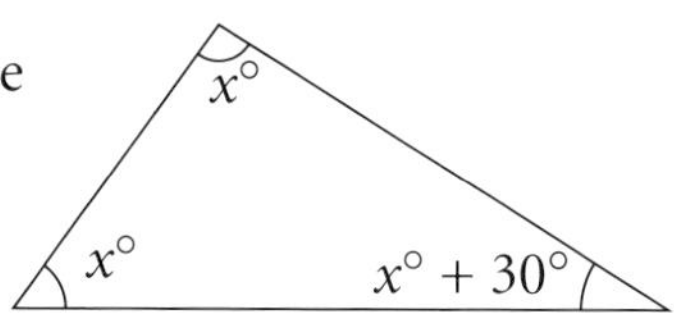

(a) Which of the following equations shows the sum of angles of this triangle? Explain your choice.

$x + 30 = 180$ $\quad$ $2x + 30 = 180$ $\quad$ $3x + 30 = 180$

$3x = 180$ $\quad$ $x + x + 30 = 180$

(b) Solve the equation that is correct.

(c) Find each angle of the triangle and check that your angles add up to 180°.

5

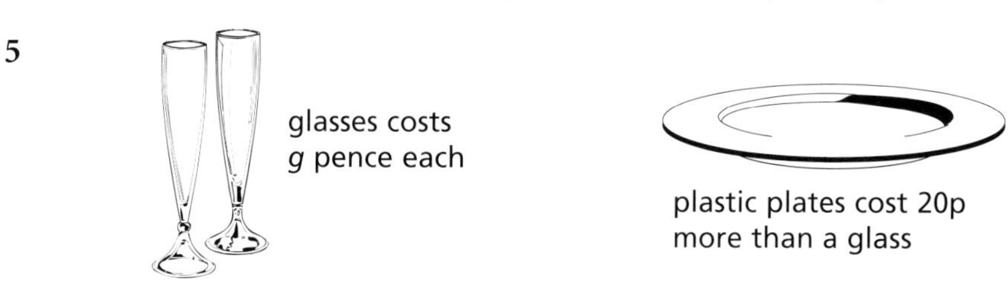

(a) Write an expression for the cost of

(i) six glasses

(ii) six plates

(iii) the total cost of six glasses and six plates.

(b) The total cost of six plates and six glasses is £5.40.
Form an equation.

(c) Solve the equation and find the cost of one glass and of one plate.

6 The combined age of three children is 38.
Amy is y years old. Zoe is three years older than Amy.
Mike is half Amy's age.
Form an equation and find each person's age.

Section C

1

$(7x + 5)$ cm

x cm | x cm

$(4x + 26)$ cm

What are the lengths of the sides of this rectangle?

2 Danny is k years old.

(a) Alice is three years younger than Danny.
How old is Alice?

(b) Danny multiplies his age by 3 and subtracts 5.
Alice multiplies her age by 4 and subtracts 8.
They both get the same answer.
How old is Danny?

3 Lauren and Faith go shopping for souvenirs.
They both start with the same number of euros.

Lauren buys 7 models of the Eiffel Tower and has 4 euros left.
Faith wants to buy 11 models but is short of 2 euros.

(a) How much does each model cost?

(b) How much money do they each have to start with?

4 Jack and Diana both think of the same number.
Jack adds 3 to his number and then multiplies by 7.
Diana multiplies her number by 5 and then adds 30.
They both end up with the same number.

Suppose the number they both think of is x.

Write down an equation in x and solve it to find the number they both thought of.

Mixed questions 2

1 Write down the equation of each of these graphs.

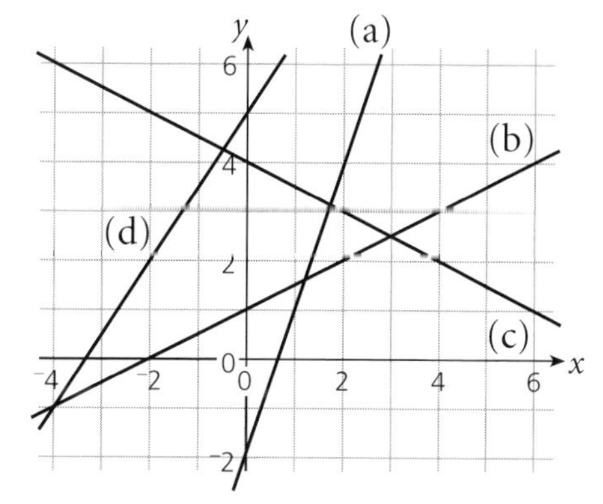

2 Describe fully the transformation that maps

(a) A to E (b) A to C

(c) C to E (d) D to C

(e) B to C (f) B to D

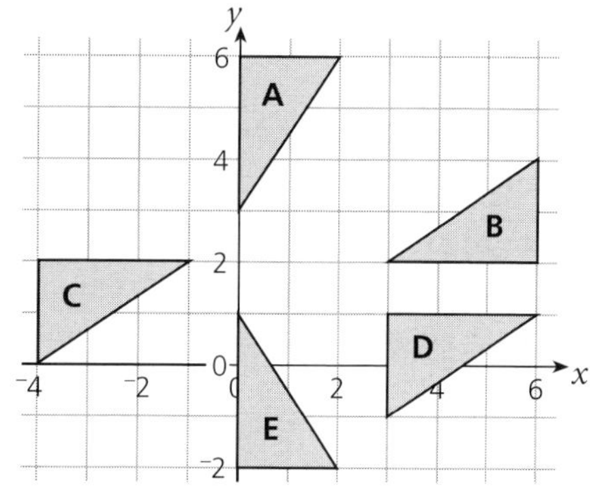

3 A train company raises its fares by 20% on 1 January. There is an outcry from the public and the company reduces fares by 15% on 1 April.

What is the overall percentage change between 31 December and 1 April?

4 Zara has invented a game for the school fair. It involves rolling a ball towards a target that moves around in an irregular way. There is no skill involved – winning is a matter of chance.

Zara tries out the game. She plays it 40 times and wins 6 times.

(a) Estimate the probability of winning.

(b) Zara charges 10p a go to play. The prize for winning is worth 50p. If 300 people play the game during the fair, how much profit would Zara expect to make?

5 Peter did a cross-country run in four sections.
The second section took 23 minutes longer than the first.
The third section took twice as long as the second.
The fourth section took 10 minutes less than the third.
Altogether the run took 3 hours 15 minutes.

How long did each section take?

6 This diagram shows the course for a sailing race.

The bearing of the first 'leg' AB is 074°.

Work out the bearing of

(a) BC (b) CA

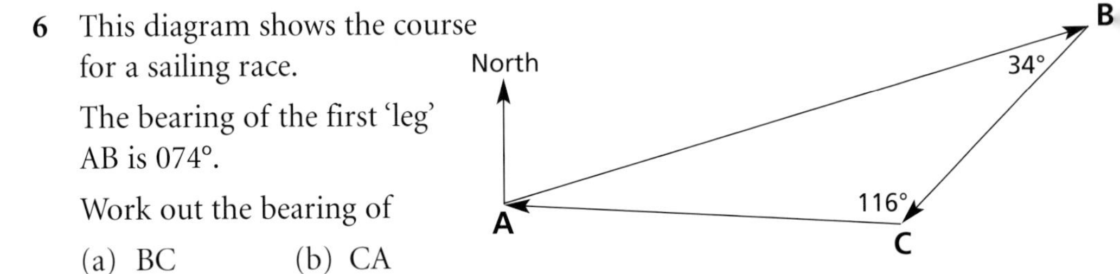

7 As a racing car goes round a track it slows down at bends.
The sharper the bend, the more it slows down.

Here are four tracks and three graphs. Each graph shows a car doing a lap round one of the tracks, passing the point A at the start and end of the lap.

(a) Match each graph to a track.

(b) Sketch a graph for the remaining track.

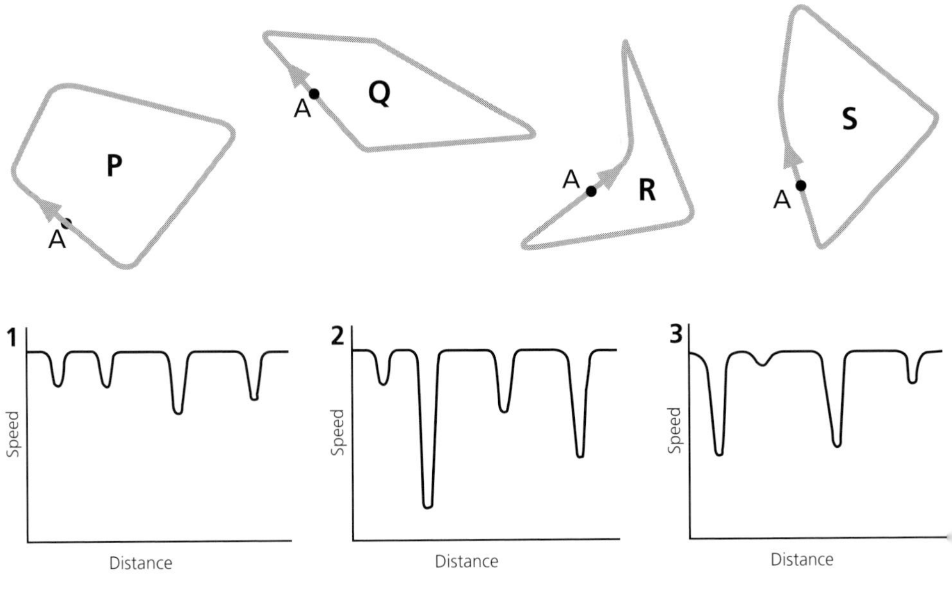

Ratio and proportion

Sections B and C

1 The time it takes Richard to type a document is proportional to the number of pages.
He takes 1 hour 20 minutes to type a 6-page document.

How long will he take to type a 9-page document?

2 The time taken to paint a ceiling is proportional to the area of the ceiling.
A ceiling of area 12 m^2 took 40 minutes to paint.

How long will it take to paint a ceiling of area 15 m^2?

3 A 2.5 litre tin of paint covers an area of 40 m^2.

(a) What area does a 1 litre tin cover?

(b) What area will 4.5 litres cover?

4 Speedipost is a parcel delivery service.
Their charges for delivering within the UK are proportional to the weight of the parcel.
A 5 kg parcel costs £20.50 to deliver.

What does a 7 kg parcel cost?

5 In May 2001, 1.39 US dollars were worth £1.
How much were 20 US dollars worth in £ and pence (to the nearest penny)?

Sections D and E

1 Meganair allows passengers to take 20 kg of baggage free. The charge for excess baggage is as shown in the graph.

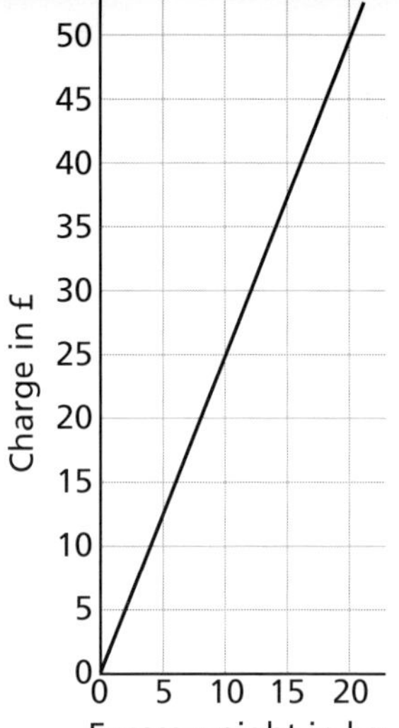

If £C is the charge for W kg of excess baggage, what is the equation connecting C and W?

2 Floor areas can be measured in square metres or square yards.
Given that 30 square yards = 25 square metres, draw a conversion graph for converting up to 60 square yards to square metres.

Use your graph to convert (to the nearest whole number)

(a) 22 square yards to square metres

(b) 14 square metres to square yards

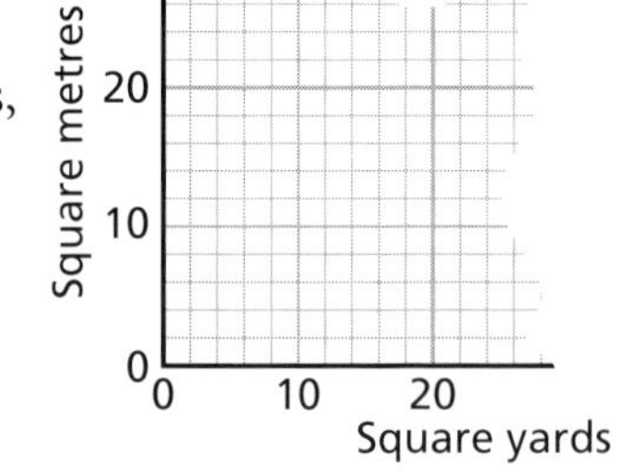

Section F

1 Sean keeps guinea pigs. The ratio of males to females is 5 : 4.

(a) What fraction goes in each of these statements?

(i) The number of females is ___ of the number of males.

(ii) The number of females is ___ of the total number of guinea pigs.

(b) If Sean has 36 guinea pigs altogether, how many of them are male?

2 Rajesh and Paul earned some money and shared it in the ratio 3 : 5.
Then Paul gave half of his share to Rajesh.

What was the ratio of their shares afterwards?

18 No chance!

Section B

1 This spinner is spun twice.

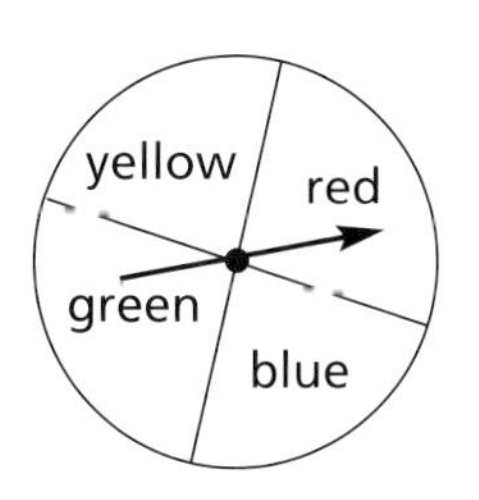

(a) List all the different pairs of colours that could be obtained.
For example,

1st spin	*2nd spin*
red	*red*

(b) What is the probability that

(i) both spins give the same colour

(ii) the two spins give different colours

(iii) neither spin gives yellow

2 These are four digit cards.

1 **4** **9** **6**

Gabbi chooses two cards randomly and places them to make a two-digit number, for example 94

9 **4**

(a) List all the possible two-digit numbers she could make. (You only use each card once!)

(b) How many different numbers could she make?

(c) What is the probability that the two-digit number that Gabbi makes is

(i) an even number

(ii) a multiple of 7

(iii) a square number

(iv) a prime number

(v) a factor of 96

(vi) not a prime number

Section C

1 For a game two spinners are used, one a square and the other a regular pentagon.

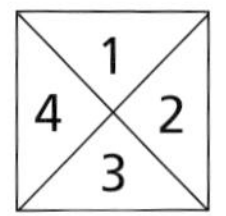

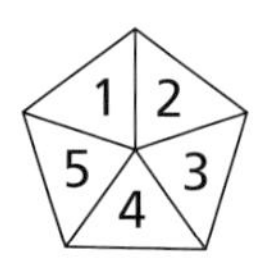

In the game the spinners are spun and the scores are added together.

Copy and complete the grid below to show all the possible totals.

Square spinner \ Pentagon spinner	**1**	**2**	**3**	**4**	**5**
1	2	3	...		
2	...	...	...		
3					
4					

2 Use your grid to find the probabilities of getting

(a) a total of 7

(b) a total of 2

(c) a total of 8 or more

(d) a double (the same number on both)

3 Which score are you most likely to get with two spinners like this?

4 In a different game two spinners like this are used

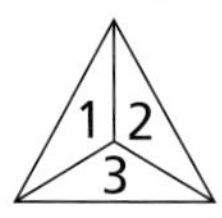

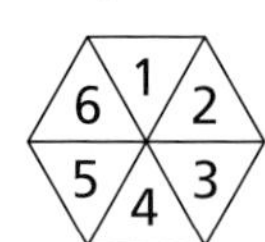

In this game the scores are **multiplied** together.

(a) Draw a grid to show all the possible scores.

(b) Use your grid to find out which score is most likely to come up.

19 Strips

Section A

1 Simplify the following expressions.

(a) $7k + k - 5k$

(b) $6 - 6m + 2 + 9m$

(c) $10 - 7n - 2 + 2n$

(d) $8 + 9p - 9 + p - 2$

(e) $6q - 1 + 5 - 10q$

(f) $15r - 7r + 8 - 8r$

(g) $7s + 4 - 4s - 3 - 1$

(h) $6t - 3 + 3t + 7 - 5t$

(i) $6u - 4 + 2u - 1$

(j) $1 + 7v - 3v - 5$

2 Simplify the following expressions.

(a) $7u + v + 8u + 2v$

(b) $2x - 4y + 9y - 4x$

(c) $2w - x - 7 + 6w - 3x$

(d) $7a + 7b - 1 + a - 11b$

(e) $6c + 8 - 3d - c - 2$

(f) $8e - 3f - 9e + 6 + f$

(g) $6g - 2 - 8g - 5h + 9$

(h) $k + 9m - 3m - 4 - 3k - 1$

(i) $2n + 3 - 3p - 5 - n + 6 - n$

(j) $q + 4 - p + 7 - q - p$

Section B

1 Copy and complete each of these addition strips.

(a)

5	8	13		

(b)

$^-2$	7				

(c)

6		9		

(d)

	7	8		

2 Copy and complete these addition strips.

(a)

p	4			

(b)

$^-5$	q			

(c)

3		r		

(d)

9		s	

3 Copy and complete these addition strips.

(a)

	q	13		

(b)

	5	s	

4 (a) Write an expression for the number in the last box in this addition strip.

5	a			

(b) What number is in the last box when $a = 2$?

(c) If the number in the last box is 37 what is the value of a?

5 Find the missing numbers in each of these addition strips.

(a)

4					52

(b)

3					31

(c)

6				0

(d)

3.2		12	

6 Find the missing numbers in these addition strips.

(a)

	5				16

(b)

		9.5					102

Section C

1 Find four pairs of equivalent expressions.

A $5x - (2x - 3)$ **B** $4x - (1 + 3x)$ **C** $2x + (x + 3)$

D $4x - (3 + x)$ **E** $2x - (x + 5)$

F $x + (2x - 3)$ **G** $2 + (x - 3)$ **H** $4x - (5 + 3x)$

2 Simplify the following expressions.

(a) $3x + (2x + 3)$ (b) $2 + (3x - 4)$ (c) $5x + (4 + 2x)$

(d) $5 - (x - 3)$ (e) $4x + (3 - x)$ (f) $7 - 2x - (4 - x)$

(g) $4 + 3x - (2 - 3x)$ (h) $3x - 6 + (5 - 2x)$ (i) $1 - x - (5x - 2)$

(j) $8 - 4x + 2x - (4 + x)$ (k) $5 - 6x - (4 - 5x)$ (l) $7 - 2x - (9 - 5x)$

Section D

1 Copy and complete each of these subtraction strips.

(a)

21	12				

(b)

24	19			

(c)

5	8			

(d)

	6	2			

2 Copy and complete these subtraction strips.

(a) | 4 | a | | |

(b) | b | 6 | | | |

(c) 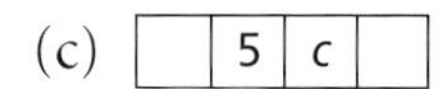
| | 5 | c | |

(d) | | d | 4 | |

(e) | e | | 2 | | | |

3 (a) Write an expression for the number in the last box in this subtraction strip.

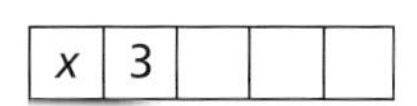
| x | 3 | | | |

(b) What number is in the last box when $x = 5$?

(c) If the number in the last box is 50 what is the value of x?

4 Work out the value of y in each of these subtraction strips.

(a) | 20 | y | | | | 5 |

(b) | y | 6 | | | 2 |

(c) | 6 | | y | | 24 |

5 Find the missing numbers in these subtraction strips.

(a) | 58 | | | 4 |

(b) | 42 | | | | | 19 |

(c) 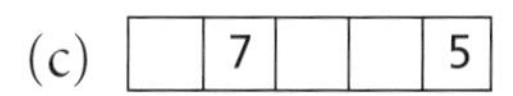
| | 7 | | | 5 |

Section E

1 Find three pairs of equivalent expressions.

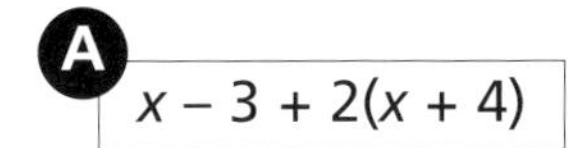

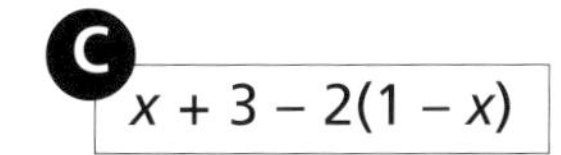

A $x - 3 + 2(x + 4)$

B $2(1 - x) - 5(1 - x)$

C $x + 3 - 2(1 - x)$

D $3 + 5x - 2(x - 1)$

E $5(x + 1) - 2(x + 2)$

F $4(x - 1) - x + 1$

2 Simplify the following expressions.

(a) $10x - 2(4x + 3)$

(b) $4y + 5(x - 4y)$

(c) $4x + 3(4 - 2x)$

(d) $10 - 4(2 + 3x) + 2x$

(e) $2x - 3 + 4(2x - 6)$

(f) $6 - x - 2(3 - 5x)$

(g) $x - 3y - 2x + 2(3y + x)$

(h) $6y - x - 2(3y - x)$

(i) $2x + 5(2x - y) + 12y$

(j) $2x - y - 4(y - x)$

20 The right connections

Section B

Here is some information on some species of whales.

Species	Mean length (metres)	Mean weight (tonnes)	Cruising speed (knots)	Number of vertebra
Great right	15.0	96.0	5	57
Bowhead	16.0	110.0	3	54
Grey	12.2	34.0	6	56
Fin	21.0	70.0	20	62
Blue	25.0	178.0	13	64
Minke	8.2	9.0	15	48
Sei	17.0	29.0	26	57
Humpback	14.6	48.0	4	53
Goosebeak	6.4	4.5	3	47
White	4.3	1.3	3	51
Great Sperm	15.0	38.0	3	40
Great Killer	8.0	6.0	5	53

1 (a) Draw a scatter diagram for the length and weight of each species.

(b) Describe any connection between the lengths and weights of these whales.

2 (a) Which whales are not very heavy compared to their lengths?

(b) What are the cruising speeds of these whales?

(c) What might be the reason why these whales are able to go at these speeds?

Sections C and D

1 A marine biologist suggests that heavier whales go slower.
She draws a scatter diagram from the data on the previous page to show this.

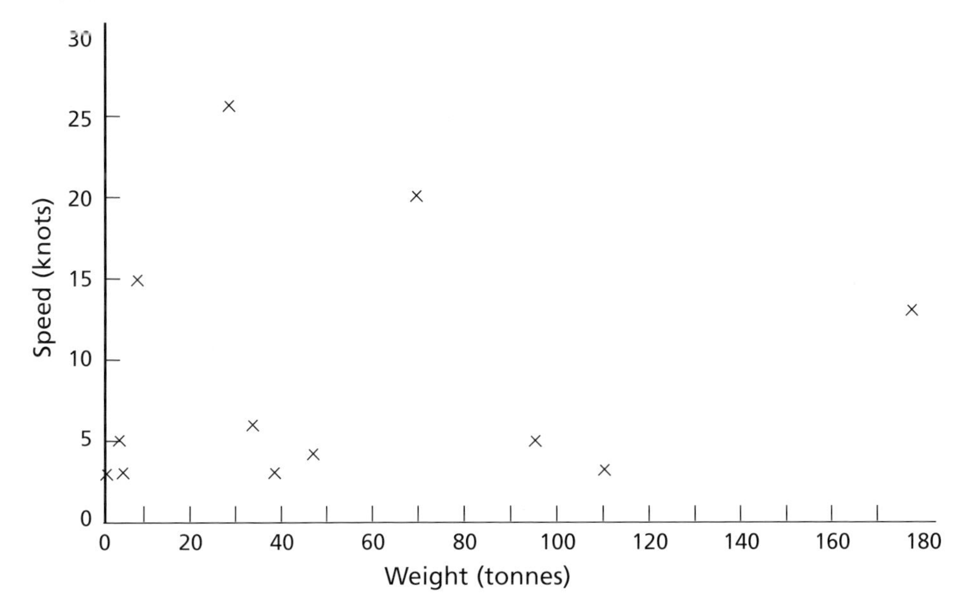

(a) Describe the correlation between weight and cruising speed.

(b) Do you agree with the marine biologist's hypothesis?

2 (a) Draw a scatter diagram of the length and the number of vertebrae.

(b) Draw a line of best fit on your scatter diagram.

(c) Use your line to estimate the number of vertebrae you would expect to find in a species of whale for which the mean length is 22 m.

3 Use the quartering method to examine the correlation between length and number of vertebrae.

21 Triangles and polygons

Section A

1 (a) Construct a triangle ABC given that
AB = 6.5 cm AC = 4.8 m angle A = 115°

(b) Measure the length BC.

2 (a) Construct two different triangles PQR given that
angle P = 40° PQ = 8.2 cm QR = 5.5 cm

(b) Measure the length of PR in each triangle.

Section B

1 Find the angles marked with letters.
Explain how you worked out each angle.

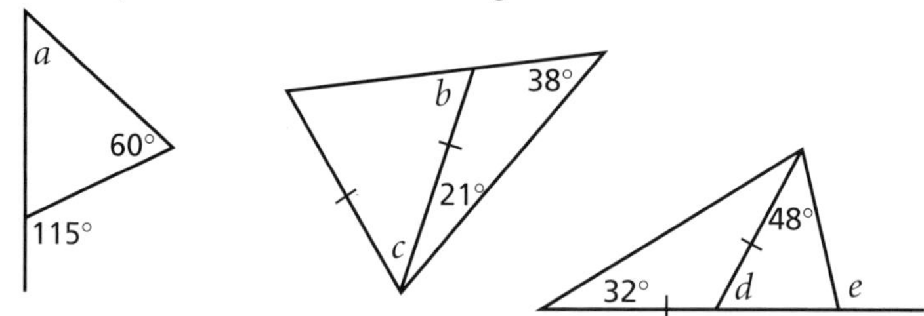

2 The angles of a triangle are $2x$, $x + 30°$ and $x - 10°$. Find the value of x.

3 Find the angles marked with letters.
Explain how you worked out each angle.

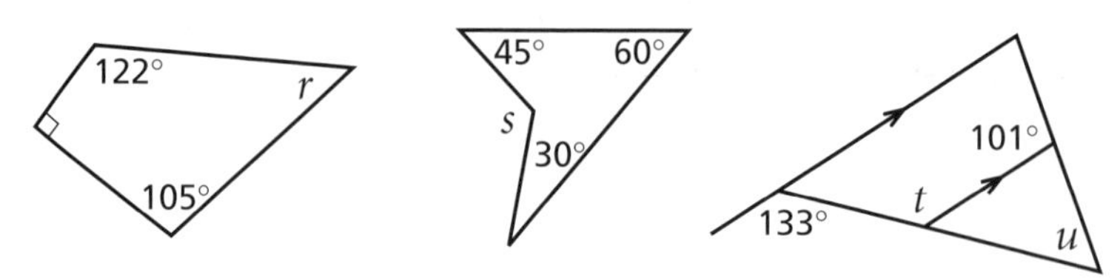

4 A quadrilateral has two right angles.
One of the remaining angles is twice as big as the other.

What are the sizes of these two angles?

Section C

1 What is the total of the interior angles of each of these?

(a) a hexagon (b) an octagon

2 Find the size of each angle marked with a letter, explaining your reasoning.

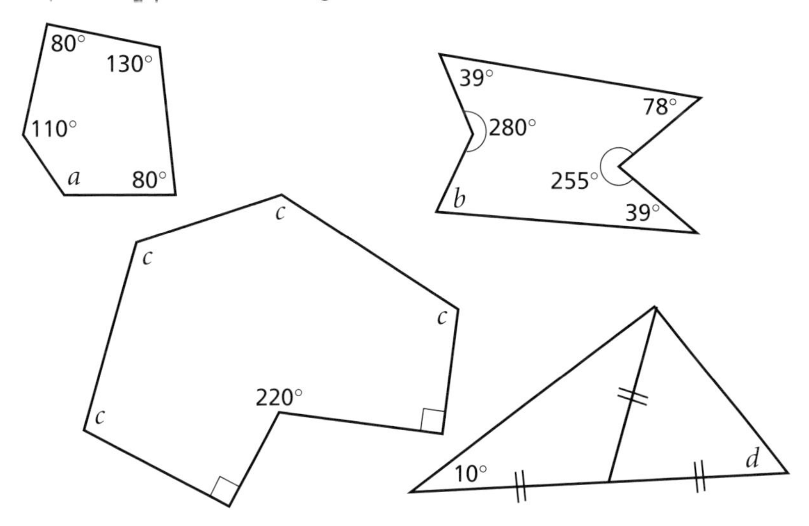

3 The total of the interior angles of a polygon is 1980°. How many sides does it have?

4 A hexagon has three right angles and the remaining three angles are equal to each other. What is the size of each of these angles?

5 In a pentagon ABCDE, angle B is 100° and angle C is 110°. The side BC is parallel to ED and the sides AE and ED are perpendicular. What is the size of angle A?

6 Find the size of angle PQS in this hexagon.

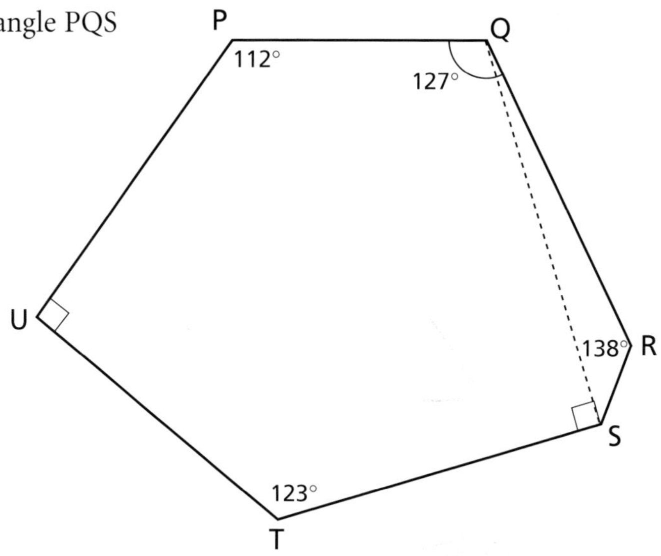

7 Find the size of the angle marked *e* in this diagram.

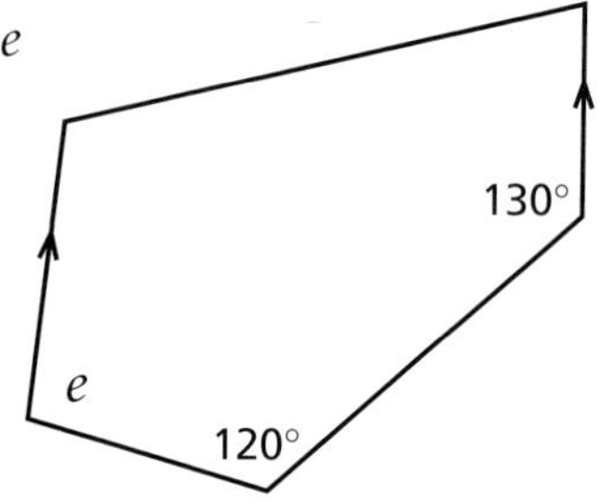

8 This polygon has a line of symmetry.
Find the sizes of the angles marked *x* and *y*.

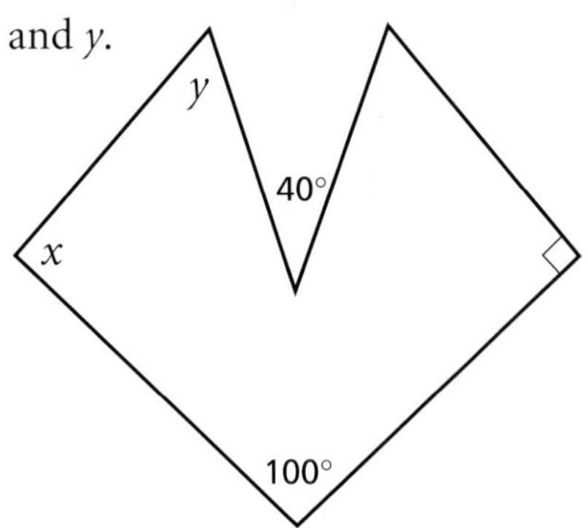

Section D

1 Find the size of each angle marked with a letter, explaining clearly how you found it.

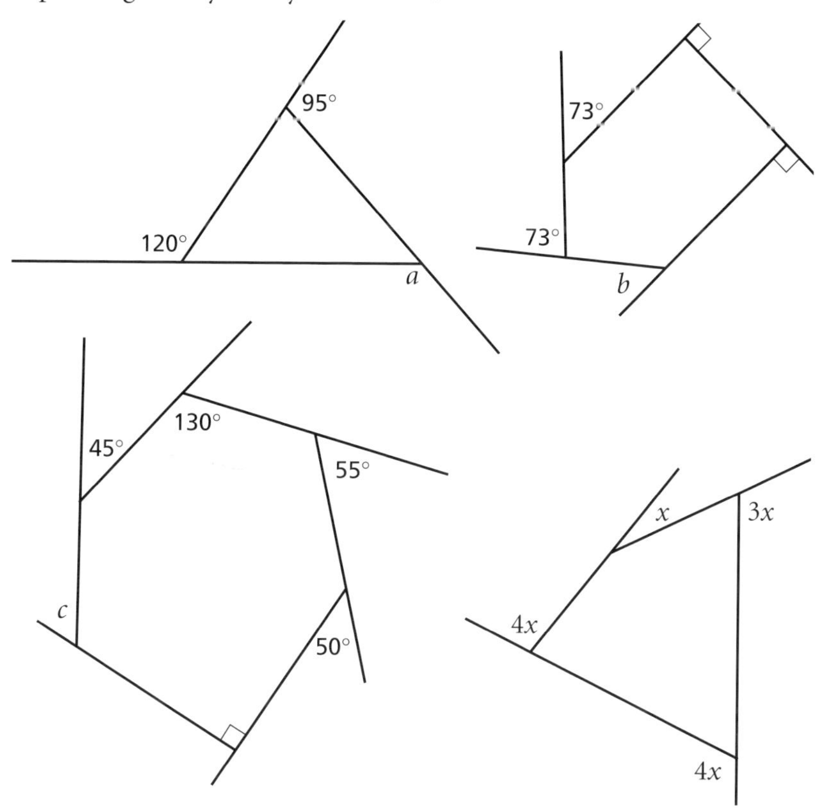

2 Three exterior angles of a polygon are 70°, 80° and 55°.
What is the total of the remaining exterior angles of the polygon?

3 What is the average size of each exterior angle of these shapes?
(a) a hexagon (b) a polygon with seven sides

4 Every exterior angle of a particular polygon is 40°.
How many sides has this polygon?

5 What is the size of each exterior angle of

(a) a regular octagon
(b) a regular dodecagon (12 sides)
(c) a regular polygon with 20 sides
(d) a regular 360-sided polygon

6 What is the size of each interior angle of each of these?

(a) a regular pentagon
(b) a regular octagon
(c) a regular dodecagon
(d) a regular polygon with 40 sides
(e) a regular polygon with 180 sides

7 These diagrams of regular polygons are not drawn to scale.
Find the size of each angle marked with a letter, explaining your reasoning.

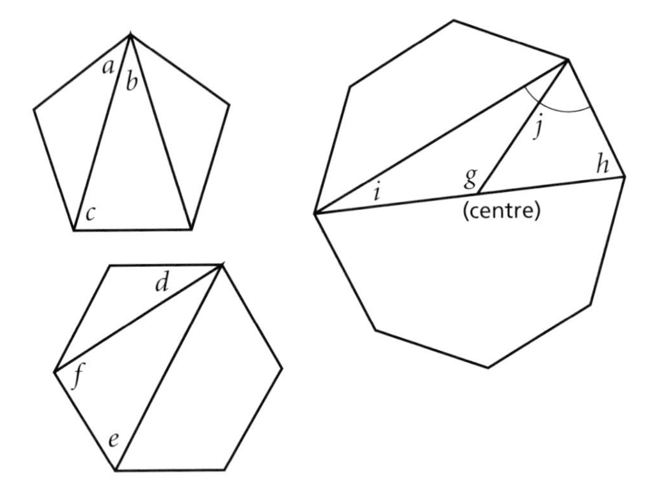

8 The diagram shows a regular octagon.
The shaded angles are exterior angles.

Find the sizes of the angles x and y, explaining clearly how you found each.

22 Moving around

Section B

1 Here is a distance table for places on the South Island of New Zealand.

Distances in km				Christchurch
			Greymouth	255
		Kaikoura	338	189
	Nelson	249	296	417
Queenstown	850	676	554	487

Here is a table showing the travelling times (by car) between the places.

Time in hours				Christchurch
			Greymouth	$3\frac{3}{4}$
		Kaikoura	5	$2\frac{1}{2}$
	Nelson	$3\frac{1}{4}$	$3\frac{3}{4}$	6
Queenstown	$12\frac{3}{4}$	$9\frac{1}{2}$	9	7

(a) Complete a similar table showing the average speed for each journey.

(b) Which journey has the fastest average speed?

(c) Which journey has the slowest average speed?

Section C

1 A train has a top speed of 125 m.p.h.
How far does it travel at top speed in (a) 3 hours (b) $\frac{1}{2}$ hour

2 A car travels at a constant speed of 60 m.p.h.
How far will it go in (a) $2\frac{1}{2}$ hours (b) $2\frac{1}{4}$ hours

3 A ship travels at a constant speed of 8 km/h.
How far does it go in (a) $5\frac{1}{2}$ hours (b) 45 minutes

4 The most economical speed to drive a car is 56 m.p.h.
If a car travels at this speed how far will it go in $\frac{3}{4}$ hour?

5 Calculate the speed, in m.p.h., of a coach that travels

(a) 105 miles in 2 hours
(b) 54 miles in $1\frac{1}{2}$ hours
(c) 8 miles in 15 minutes
(d) 4 miles in 5 minutes

Sections D and E

1 A boat travels at 12 m.p.h. How long does it take to travel

(a) 36 miles
(b) 60 miles
(c) 18 miles

2 A plane flies at 430 m.p.h.
How long, to the nearest hour, does it take to travel

(a) 5100 miles
(b) 3850 miles
(c) 1690 miles

3 (a) Change 3.14 hours to hours and minutes, to the nearest minute.
(b) Change 2 hours 52 minutes to hours, to two decimal places.

4 A train is travelling at a constant speed of 135 km/h.
How long, in hours and minutes, does it take to cover a distance of 245 km?
Give your answer to the nearest minute.

23 Substitution

Section A

1 The area of a triangle is given by the formula

$$A = \tfrac{1}{2}bh$$

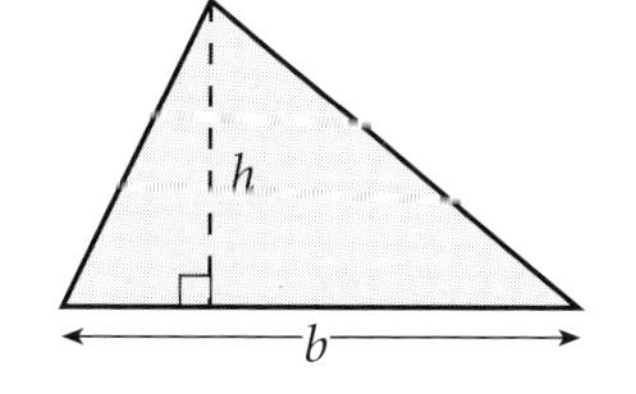

A is the area, b the base and h the height (in appropriate units).

Calculate A, stating the units, when

(a) $b = 4\,\text{cm}, h = 3.5\,\text{cm}$ (b) $b = 10\,\text{cm}, h = 8\,\text{mm}$

2 The area, $A\,\text{m}^2$, enclosed under a 'parabolic arch' is given by the formula

$$A = \frac{2wh}{3}$$

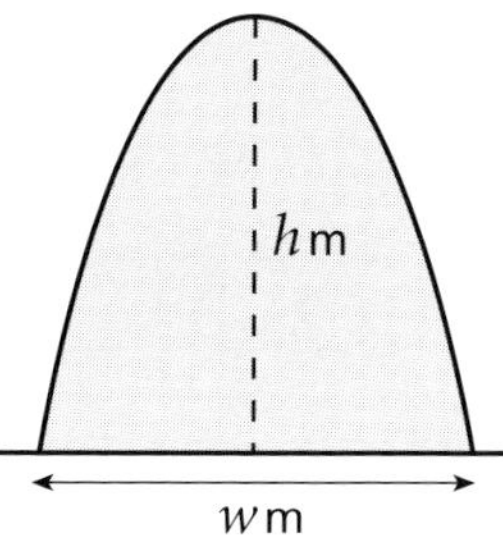

Calculate A when

(a) $w = 20, h = 60$

(b) $w = 45, h = 7.5$

Section B

1 Calculate in your head the values of each of the following when $k = 3,\ l = 5,\ m = 10$.

(a) $k(l + m)$ (b) $k(m - l)$ (c) $\frac{m + l}{k}$ (d) $2m^2$

2 Calculate in your head the values of the following when $r = 4,\ s = 2,\ t = 0.5$.

(a) $rs + t$ (b) $\frac{r}{s}$ (c) $\frac{s}{t}$ (d) $\frac{r + s}{t}$ (e) $\frac{s}{rt}$

3 If $y = 3,\ z = 4$, calculate in your head the values of

(a) y^2z (b) yz^2 (c) $(yz)^2$

All your answers should be different.

4 If $a = 5$, $b = {}^{-}2$ calculate in your head the values of

(a) $3b^2$ (b) $(3b)^2$ (c) $3(a^2 + b^2)$

5 The volume, V m^3, of a tent of the shape shown here is given by the formula

$$V = \frac{wl(r + h)}{2}$$

Calculate V when $w = 1.8$, $h = 0.6$, $r = 1.4$ and $l = 2.0$.

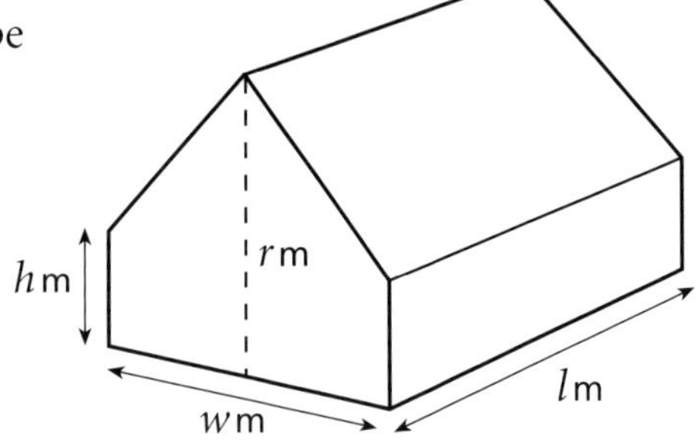

6 A small stone is thrown vertically into the air. After t seconds its height, h metres, is given by the formula $h = 50t - 5t^2$.

(a) Work out the height of the stone for the times in this table.

t	0	1	2	3	4	5	6	7	8
h									

(b) The speed of the stone, v metres per second, is given by the formula $v = 50 - 10t$.
Work out the speeds at the times in this table.

t	0	1	2	3	4	5	6	7	8
v									

(c) Interpret, in words, your answers to (a) and (b).

24 Locus

Section C

1 ABCD is a rectangle 400 m by 300 m.
There is a radio transmitter at A
and another at C.
The range of each transmitter is 300 m.

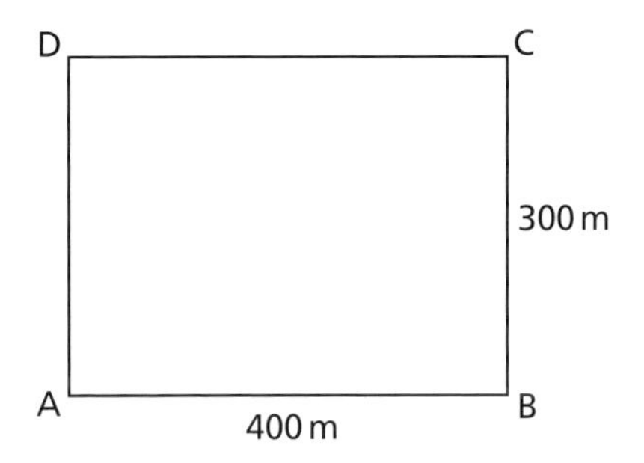

(a) Draw the rectangle to a scale of 1 cm to 100 m.
Draw the locus of all points inside the rectangle which are 300 m from A.

(b) Draw the locus of all points inside the rectangle which are 300 m from C.

(c) Shade the locus of all points inside the rectangle that are within range of both transmitters.

2 (a) Draw this right-angled triangle.
Draw the locus of all the points inside the triangle that are equidistant from the sides AB and AC.

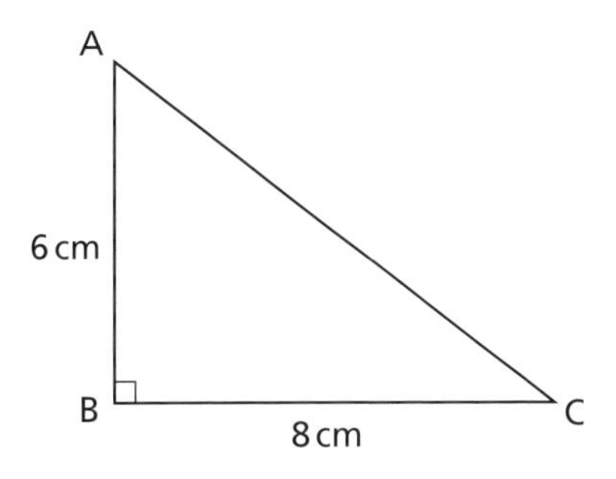

(b) The locus you have drawn meets BC at the point X.
Measure BX and check that the distance from X to AC is equal to BX.

Section D

1 The diagram shows the plan of a shed against the wall of a house.
A goat is tethered to the point A by a rope of length 7 m.
The goat starts at the point G_1 and walks so that the rope is always taut.
G_2 shows a later position of the goat.

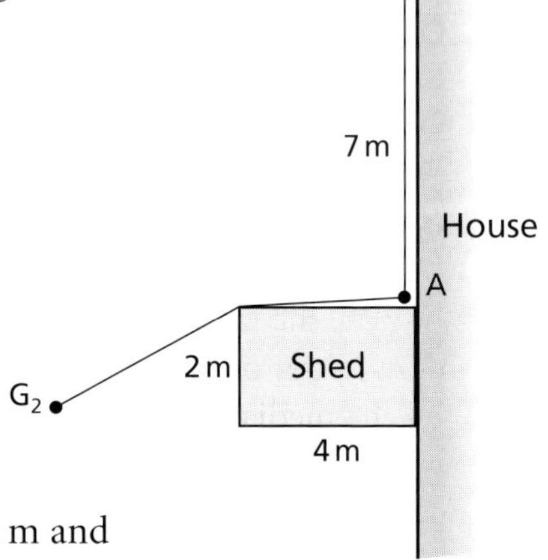

Draw the diagram to a scale of 1 cm to 1 m and draw the locus of the walking goat.

2 A square with sides 2 cm long slides at a constant speed from position 1 to position 2 as shown below.
An insect crawls at a constant speed along the diagonal of the square, starting at A and finishing at B.
Draw the locus of the insect.

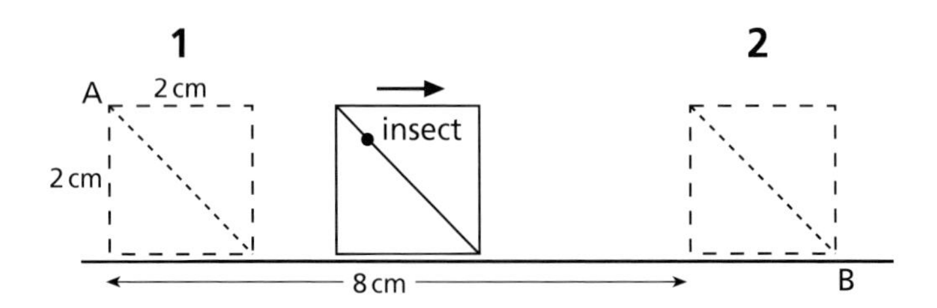

25 Distributions

Section A

1 Here are the weights in grams of a litter of baby mice.

19, 18, 16, 20, 17, 16, 19

(a) Find the median weight.

(b) Find the range of the weights.

2 These are the tail lengths in mm of another litter of mice.
Find the median and range.

46, 58, 57, 48, 55, 51, 49

3 Zoe keeps lots of mice.
Here are her records for the numbers of mice born in each litter.

6, 5, 7, 5, 4, 5, 6, 7, 6, 7, 5, 8

(a) What was the mode for the number of baby mice?

(b) What was the median number of baby mice?

4 Sam also keeps lots of mice.
These are the records of the number of mice born in each litter over the last few months.

Number of mice in litter	Tally	Frequency
4	//	2
5	~~////~~	5
6	~~////~~ /	6
7	~~////~~ ///	8
8	///	3
9	/	1

(a) How many litters of mice are there?

(b) What is the range of the number of mice in a litter?

(c) What is the mode for the litter size?

(d) What is the median litter size?

5 Look at your results for questions 3 and 4.
Use them to compare litter sizes for Zoe's mice and Sam's mice.

Section B

1 Here is a stem-and-leaf table showing the length of time, in minutes, that it took some pupils to solve a mathematical puzzle.

Stem	Leaf
0	4 6 8
1	0 2 3 5 5 7
2	1 3 4 6 8 8 9 9
3	0 1 2 2 5
4	2 6 7 9
5	5

(a) Find the range of the times.

(b) Find the number of people who completed the puzzle.

(c) Find the median time.

(d) Find the modal group.

2 In a biology experiment 25 pupils were growing sunflowers in two different types of soil. Here are the heights of their plants in cm after one month.

Soil A

26, 30, 48, 59, 62, 25, 34, 36, 41, 43, 47, 56, 51, 45, 24, 32, 38, 32, 41, 40, 57, 60, 21, 39, 47

Soil B

29, 33, 44, 48, 51, 53, 35, 42, 50, 46, 48, 26, 29, 33, 51, 30, 28, 41, 47, 54, 50, 42, 35, 46, 48

(a) For each type of soil

(i) make a stem-and-leaf table

(ii) find the range of heights

(iii) find the median height

(iv) find the modal group

(b) Use these results to compare the two soils.

3 24 pupils weighed their school bags. Here are the weights in kg.

2.4	3.1	1.9	2.8	3.4	4.2	4.0	4.4
3.1	2.6	2.7	3.9	1.8	2.0	1.5	3.4
4.3	2.6	2.9	3.2	2.2	3.7	4.1	4.2

(a) Make a stem-and-leaf table to show these weights.

(b) What is the range of the weights?

(c) What is the median weight?

Section C

1 Anna has the results of five of her tests: 48% 62% 59% 71% 68%

(a) What is her mean score?

(b) She receives her last result, 59%. What is her new mean score?

2 Each pupil in a class counted how many words they had written on one side of a piece of A4 paper. Here is the data they collected.

Girls	304	328	118	247	306	120	218	125	272	228	279	335		
Boys	348	206	116	259	212	186	315	107	164	319	243	96	137	311

(a) Find the mean and range for the girls and boys separately.

(b) Write two sentences comparing the number of words for girls and boys.

(c) Find the mean number of words for the whole class.

3 Sunil checked the contents of some boxes of nails. His results are shown in this frequency table.

Number of nails in the box	Frequency
48	4
49	6
50	12
51	8

Calculate

(a) the number of boxes Sunil checked

(b) the total number of nails in all the boxes

(c) the mean number of nails per box

4

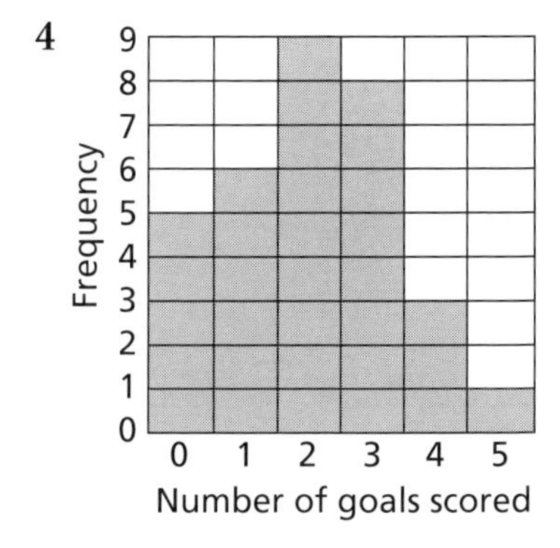

Rob kept a record of the number of goals scored per match for his local football league. His results are shown in the bar chart.

Calculate

(a) the number of matches that were played

(b) the total number of goals that were scored

(c) the mean number of goals per match, correct to 1 d.p.

Sections D and E

1 Sasha conducted a survey in her class to find out how many pens, pencils and other writing implements each pupil had.
Her sister Hadeel did a similar survey in her class. Here are their results.

Sasha's class

4	2	8	7	9	12	15	7	5	4	3	6	4
2	1	12	14	17	15	7	18	1	9			

Hadeel's class

12	15	4	7	9	10	14	16	2	3	7	10	8
12	17	15	16	20	9	4	15	17	19			

(a) Make a grouped frequency table for each class.
Use groups 1–5, 6–10 and so on.

(b) Draw a frequency bar chart for each class.

(c) What is the modal group for Sasha's class?

(d) What is the modal group for Hadeel's class?

2 Here are the heights, in metres, of the pupils in a Year 9 class.

Boys	1.42	1.35	1.53	1.47	1.45	1.39	1.51	1.56	1.48
	1.36	1.28	1.51	1.30	1.34	1.47			
Girls	1.35	1.55	1.40	1.29	1.32	1.52	1.41	1.26	1.35
	1.28	1.44	1.38	1.39	1.30	1.43			

(a) Make grouped frequency tables for boys and girls.
Use intervals 1.20–1.30, 1.30–1.40 etc.
(Make sure you know which interval 1.30 and so on will go into.)

(b) Draw frequency polygons for boys and for girls on the same axes.

(c) Write a sentence or two comparing the two sets of heights.

3 This chart shows the ages of people using a library one day.

(a) Which is the modal interval?

(b) How many people under 20 use the library?

(c) What can you say about the age of the oldest user of the library?

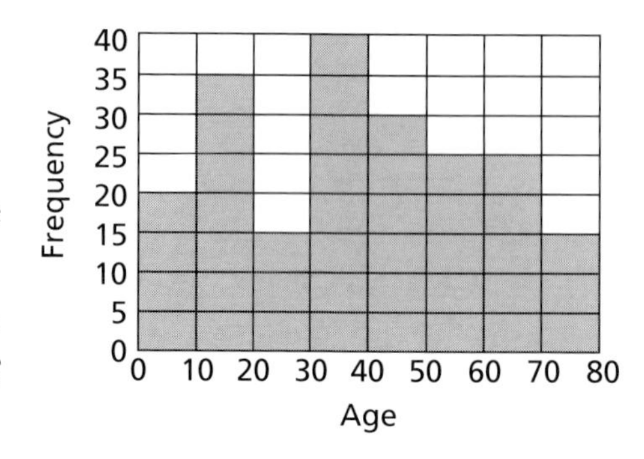

Section F

1 (a) Copy and complete this table showing the weights, in kg, of some athletes.

Weight in kg	Mid-interval value	Frequency	Mid-interval value × frequency
50–60		8	
60–70		16	
70–80		10	
80–90		6	
	Totals		

(b) Calculate an estimate of the mean weight for these athletes.

2 Use this data to calculate an estimate of the mean journey time.

Journey time in minutes	Frequency
0–20	6
20–40	15
40–60	23
60–80	21
80–100	12

3 The table gives the lengths of the leaves, in cm, of some plants. Calculate an estimate of the mean leaf length for these plants.

Leaf length	0–5	5–10	10–15	15–20
Frequency	22	35	29	14

4 This frequency polygon shows the ages of people visiting a cinema.

(a) What is the modal age group?

(b) How many people are included in the survey?

(c) Draw up a frequency table showing the data from the frequency polygon, and hence calculate an estimate of the mean age of the cinema goers.

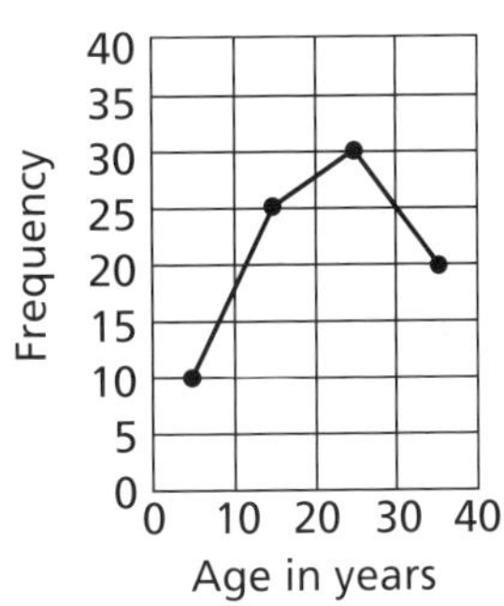

Section G

1 These are the ages of the people on a coach trip.

6	6	7	7	7	8	8	9	9	9	9	10
11	11	11	12	12	12	13	13	14	32	35	44

(a) What is the modal age of the people on the coach trip?

(b) What is their median age?

(c) Calculate the mean age.

(d) Which average, if any, do you think gives the best representation of the age of the people on the coach?
Give a reason for your answer.

2 The following donations were put into a collecting box by a party of visitors.

2p	5p	5p	10p	10p	10p	10p
10p	20p	20p	50p	50p	£1	£1
£1	£1	£5	£10			

(a) Find the mean, median and mode for the donations.

(b) Which, if any, do you think gives a typical value for a donation?
Explain your answer.

3 The heights of some British and some Japanese students of the same age are to be compared.

British students – heights in cm			
154	145	142	156
149	152	150	138
145	158	139	140

Japanese students – heights in cm			
134	142	140	153
141	137	151	137
142	147	140	143

(a) Find the range of heights for the British students and for the Japanese students.

(b) Find the median and the mean for both sets of data.

(c) Write a sentence or two comparing the heights of the students.

Mixed questions 3

1 Nita lives 126 miles from London.
She drove from home to London in $3\frac{1}{2}$ hours but took $4\frac{1}{2}$ hours for the return journey.

(a) Write the ratio *time to London : time from London* in its simplest form.

(b) Find the ratio *average speed to London : average speed from London* and write it in its simplest form. What do you notice?

2 Lewis has a 30-sided dice with faces numbered 1 to 30 and an ordinary dice.
He rolls both of them.

(a) How many equally likely outcomes are there for the pair of rolls?

(b) List all the outcomes for which the number on the ordinary dice is greater than the number on the 30-sided dice.

(c) What is the probability that the number on the ordinary dice is greater than the number on the 30-sided dice?

3 An irregular hexagon has three angles of size x and three angles of size $2x$.
Find the value of x.

4 Find the value of

(a) $u^2 + 2v^2$ when $u = 3,\ v = 2$ (b) $a - \frac{b}{c}$ when $a = 5,\ b = 3,\ c = \frac{1}{2}$

5 ABCD is a parallelogram.

(a) Draw the parallelogram and construct the locus of all points that are equidistant from A and C.

(b) Mark the point on AB that is equidistant from A and C.

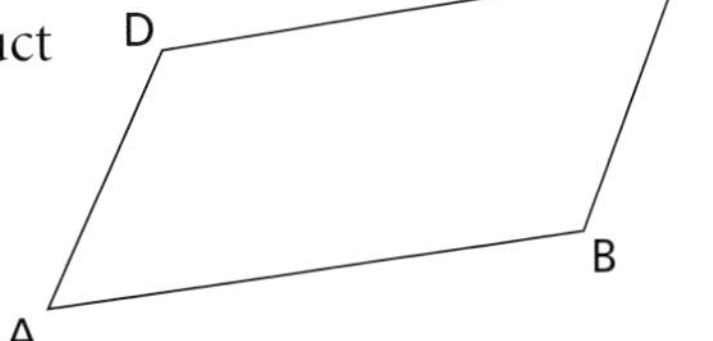

6 In a school there are four classes of 27 pupils, three classes of 28 and two classes of 30. There are 12 teachers in the school.

(a) Calculate the mean class size.

(b) Calculate the ratio of pupils to teachers in its simplest form.

7 A train leaves Birmingham at 0745 and arrives in London at 0905.
The distance from Birmingham to London is 102 miles.
Calculate the average speed of the train, to the nearest 0.1 m.p.h.

8 Fifteen students took tests in French and maths.
The marks (out of 20) are shown in this table.

Student	A	B	C	D	E	F	G	H	I	J	K	L	M	N	[illegible]
Mark in French	8	11	13	9	15	10	8	15	6	7	11	17	9	12	1
Mark in maths	4	16	8	18	6	8	13	12	7	15	7	10	6	11	1

(a) Draw a scatter diagram to show this data.

(b) What does the scatter diagram tell you?

9 Danny's uncle gave him some money on his tenth birthday.
On his eleventh birthday his uncle gave him £10 more than on his tenth birthday.
On his twelfth, thirteenth, fourteenth and fifteenth birthdays, each time Danny got £10 more than on the previous birthday.
On his sixteenth birthday Danny's uncle gave him twice as much as on his fifteenth birthday.

Altogether, Danny's uncle gave Danny £470.
How much did he give Danny on his tenth birthday?

10 This fair dice, with faces numbered 1 to 5, is rolled twice.
Find the probability that the total of the two scores is an even number.

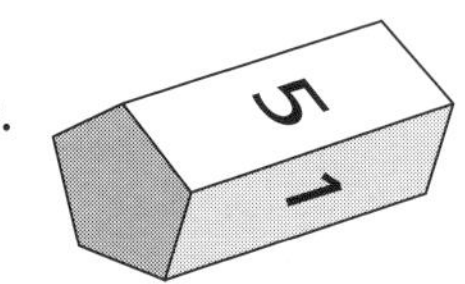

11 A shop gives its prices in £ and in euros (€).
Find the missing prices on these price tickets.

12 This frequency polygon shows the distribution of the lengths of a collection of snakes.
Calculate an estimate of the mean length.

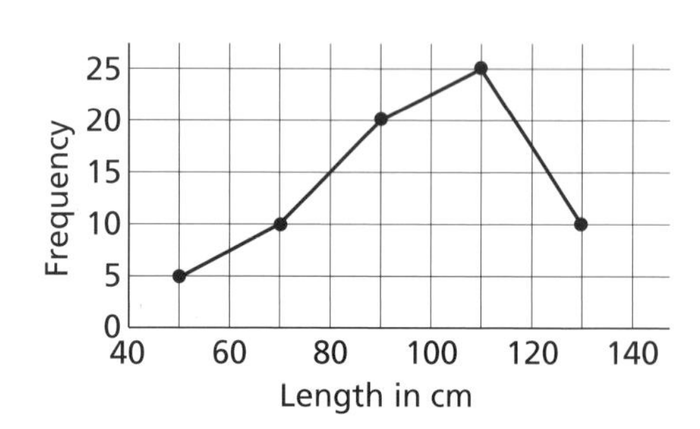